Fuels and Petrochemicals Division 2015

Core Programming Area at the 2015 AIChE Spring Meeting and 11th Global Congress on Process Safety

Austin, Texas, USA
26-30 April 2015

ISBN: 978-1-5108-0703-7

Printed from e-media with permission by:

Curran Associates, Inc.
57 Morehouse Lane
Red Hook, NY 12571

CURRAN ASSOCIATES INC. proceedings.com

Printed by Curran Associates, Inc. (2015)

AIChE
3 Park Avenue
New York, NY 10016-5991

Phone: (203) 702-7660
Fax: (203) 775-5177

www.aiche.org

Additional copies of this publication are available from:

Curran Associates, Inc.
57 Morehouse Lane
Red Hook, NY 12571 USA
Phone: 845-758-0400
Fax: 845-758-2634
Email: curran@proceedings.com
Web: www.proceedings.com

TABLE OF CONTENTS

399812 Condensed Phase Ketonization of Organic Acids Produced By the Hydrothermal Liquefaction of Lignocellulosic Biomass

Abstract:

The direct liquefaction of lignocellulosic biomass (e.g. fast pyrolysis, catalytic fast pyrolysis, or hydrothermal liquefaction) and subsequent bio-oil upgrading is one strategy for producing hydrocarbon fuels from biomass. The aqueous stream generated during the hydrothermal liquefaction (HTL) of terrestrial biomass contains 35-40 percent of the total carbon fed to the HTL reactor. Short chain organic acids (e.g. acetic acid) are one of the primary organic constituents of the aqueous phase. The ketonization of these short organic acids whereby two organic acid equivalents react to form a single ketone in the condensed aqueous phase was investigated. Ketonization is desirable because 1) a C-C bond is formed between two shorter chain compounds; 2) ¾ of the O present in the original reactants is rejected as one equivalent of CO_2 and H_2O each without the need for additional hydrogen; 3) the ketone products have higher vapor pressure than the parent acids and can be readily concentrated via distillation. Furthermore, performing the reaction in the condensed phase is advantageous because the energy intensive step of vaporizing a high volume stream rich in water is avoided. High throughput catalyst screening via small (~4 mL) batch reactors at 275°C and ~1000 psig (N_2) was performed to identify active catalysts. Selected catalysts were tested in a 3 mL flooded bed continuous reactor at 300°C and 1350-1400 psig. The effect of WHSV and acetic acid concentration on the rate of converting acetic acid to acetone was investigated. Model compounds as well as real feeds were tested. Under similar conditions, the rate of ketonization using the real feed was only about one-half that observed with model compounds. The lowered rate when utilizing real feeds may be due to a high concentration of solubilized Na (4000-6000 ppm) present from the HTL process. Fresh and spent catalyst characterization will be reported to suggest catalyst structure/activity relationships. Downstream, it is envisioned the ketones could undergo further processing such as reduction and dehydration steps to produce olefins. An overall flow diagram of the conceptualized process for producing olefins as well as preliminary techno-economic analysis incorporating an HTL bio-refinery will be presented.

400046 Thermogravimetric Analysis of Bio-Oils and Upgraded Products

Abstract:

Biomass can be converted to liquid products through thermochemical processing such as pyrolysis and hydrothermal liquefaction. This bio-oil, after catalytic upgrading to remove reactive oxygenated species, can be used to produce fuel products and commodity chemicals. A key metric for fuel product composition and quality is the boiling point range or distillation curve. Determining the applicability of bio-oil and upgraded products as fuels requires knowledge of distillation curves. Standard distillation methods used to characterize petroleum products can be applied to bio-derived fuels; however, these methods require relatively large sample volumes that may not be available from lab scale research reactors. Although gas chromatography (GC) based simulated distillation methods that require small sample volumes exist for petroleum products, these are not applicable to bio-oils containing oxygenated compounds. Thermogravimetry (TG) is well suited for determining boiling temperatures of oxygenated bio-oils using small sample sizes. TG based micro-distillation data indicate bio-oil reactivity at high temperatures and thus provide a rapid determination of the extent of removal of reactive species. This paper discusses TG micro-distillation methodology for bio-oils and comparison to GC based simulated distillation. Both methods are applied to pyrolysis and hydrothermal liquefaction products as well as upgraded products with three levels of oxygen removal. These data highlight the impact of oxygenated compounds on GC simulated distillation data as well as the use of TG data to indicate removal of reactive compounds.

400086 Opportunities for Co-Processing Biomass Derived Intermediates in a Petroleum Refinery

Abstract:

As the biofuels industry broadens to include advanced liquid hydrocarbon fuels, understanding commercialization opportunities and leveraging existing infrastructure to reduce costs will become critical for the nationwide deployment. From the bio-intermediates producer's perspective, use of existing infrastructure may reduce investment costs. From a refiner's perspective, there is available refining capacity that could be utilized for upgrading bio-intermediates. However there are numerous challenges associated with refinery integration, such as the uncertainties associated with the technical risks. Additionally, there are no clear economic incentives as Renewable Identification Number (RIN) credits are not yet allowed under the Renewable Fuel Standard (RFS) for co-processing. Per the 2014 EIA Annual Energy Outlook Report, the transition of the fuel pool to a higher diesel and lower gasoline demand has a large impact on petroleum refinery operations. Specifically FCC utilization is expected to decrease from 67% in 2012 to 55% by 2040 due to lower gasoline demand. This allows for the opportunity to supplement this processing availability by using biomass derived intermediates, particularly if diesel fuel is the primary target from these biomass products. Alternatively, hydrocracking demand is expected to grow with capacity being increased by 30% from 2012 to 2040. With increasing hydrocracking capacity, biomass intermediates could potentially supplement the petroleum feedstock for these units. A study of biomass availability around US refineries, and a preliminary assessment of the impact of co-refining were presented previously. As a continuation of that work, the team leveraged existing experimental data, and modeling expertise to develop first of kind Aspen models for co-processing bio-derived intermediates in a petroleum hydrocracker and fluidized catalytic cracking unit to assist with this assessment. Sensitivity analysis, both with and without the addition of a bio-oil intermediate feed was used to qualitatively and quantitatively address co-processing impacts. The preliminary impact of refinery integration of biomass derived intermediates will be presented.

400978 Direct Conversion of Mixed Oxygenates to Isobutene over $Zn_xZr_yO_z$ Mixed Oxides

Abstract:

Direct conversion of ethanol to isobutene has recently been reported over ZnxZryOz mixed oxide catalysts with balanced acid-base sites [1]. A complex catalytic reaction pathway to hydrocarbons starting from ethanol occurs through a combination of reactions, including alcohol dehydration, aldol condensation, dehydrogenation, and Bronsted acid-site catalyzed acetone-to-isobutene. At Pacific Northwest National Laboratory we have been exploring the use of this catalyst system for the direction conversion of a variety of oxygenated species, including acetic acid, acetaldehyde, and ethyl acetate. This presentation will discuss reaction mechanism and application for the direct conversion of complex oxygenated feed mixtures to isobutene over ZnxZryOz mixed oxide catalysts.

[1] Sun et al. 2011. J Am Chem Soc. 133: 11096-11099

398516 Kraft Lignin Depolymerization Under Hydrothermal Conditions

Abstract:

Kraft lignin depolymerization was conducted in water and mixture of water and alcohols with different ratios. The highest lignin conversion yields obtained was 82%. Direct quantitative analysis of the depolymerized products by HPLC was completed without organic solvent extraction. The yields for six standard compounds were obtained from the HPLC analyses. The paper also compared the yields of lignin depolymerization from hydrothermal processing and pyrolysis.

399535 Qualitative and Quantitative Analysis of Oxygenates in Oils Derived from Biomass

Abstract:

The production of liquid transportation fuels from biomass in an efficient and cost-effective way is a primary focus of the US DOE. Conversion of wood and other agricultural residual feedstocks through a liquefaction method, such as fast pyrolysis or hydrothermal liquefaction, followed by a catalytic upgrading pathway is one of the most promising ways to produce fuels and chemicals. The resultant oil is a complex liquid mixture of paraffins, aromatics, oxygenates and other components. It goes without saying that a near complete analysis of these oils will aid the subsequent processing steps. Main differences between bio-derived oils and petroleum based oils include the high water, acid, and oxygen content of the former. Using current technologies we are able to characterize and account for the majority of the carbon containing species, but still striving to account for 35% unknowns. We also believe that oxygenates present in these oils results in instability and cause complications in down-stream processing. A better knowledge of oxygenates present in bio-oil is needed to properly evaluate bio-oil quality and may provide the key to solving processing issues encountered in the upgrading and refining. In this current climate of studying the possibility of integrating bio-oils as a co-feed in intermediate refinery streams, an intimate knowledge of species present in bio-oils and their abundance is a key research need. We are working towards standardizing analytical methods to identify and quantify species such as carbonyls, carboxylates, phenolic, and hydroxyl species in bio-oils. Our work in this regard will be discussed in this paper.

Abstract Number 399548
Extended Abstract

Harold N. Knickle, Chemical Engineering, University of Rhode Island, Kingston, RI, Oleg Kravchenko, Podgorny Institute, Kharkov, Ukraine, Igor Baranov, Podgorny Institute and Iryna Suvorova, IPMach, Kharkov, Ukraine

Abstract

In some countries limited reserves of liquid fuel as well as the general trend to reduce atmospheric contamination poses a challenging task for producers of heat energy and electricity. Solutions include methods to increase fuel combustion efficiency and at the same time reduce the amount of emissions. The increase in fuel combustion efficiency as a solution to this problem resides in continuously improving the design of burners and the technology of liquid fuel combustion. An increase in liquid fuel can be accomplished by treatment by cavitation of mixtures of black oil, water, unconditioned hydrocarbons, refinery and coal industry wastes, tank washes, biomass and other materials. Black oil contains hydrocarbons with molecular weights from 70-210.

Flow in intricate channels of fluid vortex converters is modelled both mathematically and experimentally. The key task was to obtain ultra-fine emulsions and suspensions. The maximum effect is achieved with fluid dynamic and acoustic cavitation in the flow fuel mixtures. Results for experiments in obtaining artificial composite liquid fuel (ACLF) and their effective combustion with the cavitation devices developed are given in this paper. The cost effectiveness of using these developments and the methods of implementing the results in different industries were demonstrated.

Keywords: mathematical model, cavitation nozzle, liquid fuel

Review of recent studies and publications.

As practice has shown, the effectiveness of fuel combustion largely depends on how fine the atomizing is. The most promising method of atomizing liquid fuel is by using dedicated nozzle cavitators that ensure continuity disruption and improve the dispersion of liquid fuel [1].

Disruption of droplet fluid continuity under the effect of tensile stresses occurs when rarefaction in a confined space of the fluid forms voids – cavity pockets or cavitation bubbles – filled with fluid vapor and dissolved in product gases. Cavitation bubbles and cavity pockets are formed in sites where fluid pressure is lower than the critical one. These phenomena involve secondary effects in the fluid that intensify mixing, fragmentation and dispersion of liquid components and suspended solids in the fluid mixtures.

The key trend in developing advanced burners is the improvement in atomizing fineness to increase the fuel and oxidizer reaction surface. The authors have implemented such a process by developing new types of burners.

Unsolved part of the general problem.

Under scarce gas conditions and transition to liquid fuel it is necessary to identify more effective and advanced methods for obtaining the combustion of liquid fuel. This is especially important when combusting alternative kinds of fuel, since commercial nozzles fail to ensure required droplet fineness, and hence, do not meet the above requirements.

Objective.

The objective of this study is developing a nonconventional approach is the creation of effective devices for fluid dynamic action, with a goal of qualitatively improving the energy and environmental indicators of current technologies and of obtaining and combusting liquid fuel.

Nozzle Design –Cavitation Fluid-Vortex Converters

The A.N. Podgorny Institute for Mechanical Engineering Problems NAS of Ukraine has developed new unique types of cavitation fluid-vortex converters (FVC). They are used effectively in the oil and oil product cracking process, and as nozzles in burners [1].

The theoretical investigations of processes to design cavitation nozzles are based on mathematical modelling of fluid flow in intricate channels. Numerical solutions were found using the R-functions method [2,3]. Different channel flow conditions were investigated with the ***POLE*** software package developed under supervision of the late Academician V.L. Rvachev in the Department for Applied Mathematics and Computational Methods [4] in our laboratory. In the Pole software system the R-function method with variation methods (Ritz method, Bubnov-Galerkin method, least squares method) are used for solving boundary value problems..
Flow patterns in intricate channels were obtained using the POLE software (Fig. 1.) [2,3].

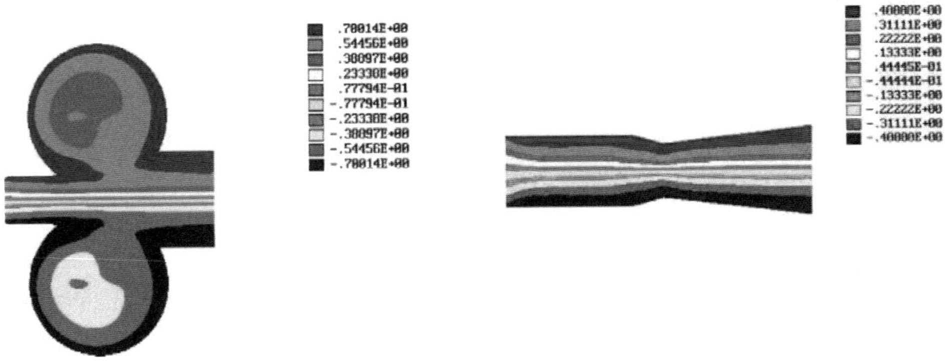

Fig. 1. Flow functions in channels

Designs of cavitation devices were obtained based on numerical investigations. Figure 2 is one cavitation model.

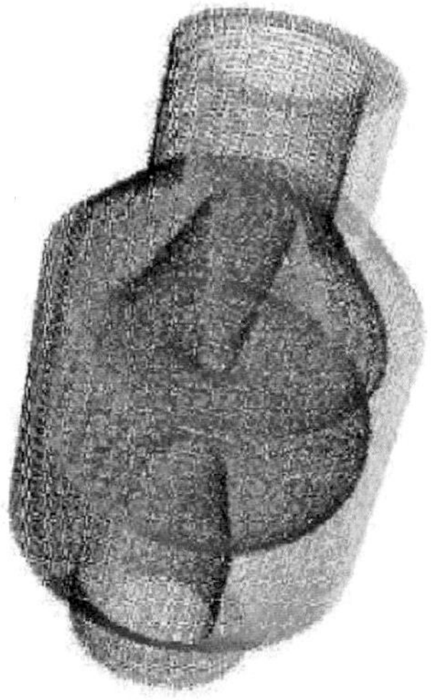

Fig. 2. Cavitation nozzle model

Figure 3 depicts the cavitation nozzle. The flowing part of the nozzle is an axisymmetric channel whose inlet has convergent tube 2 and 4, and the outlet has diffuser 5. The toroidal vortex-mixing chamber 6 is located in the midsection of the channel (Fig. 3). Because rarefaction occurs in chamber 6 during fluid flow in the channel, the second fuel component is sucked in via ejection orifices 8 from auxiliary annular chamber 7. Ejection channels 8 communicate with the rarefaction zone of toroidal chamber 6. The ejected flow sucked in from chamber 7, is separated into *m* individual orthogonal flows. Each of the *m* flows is subject to torsion moments created by the shape of the ejection orifices. The number *m* of ejection channels, and their shape and location in the construction is defined with a technique developed by the authors.

Nozzle Diffuser Channel Operation

In the diffuser channel, cavitation proceeds further; the combustion mixture film is broken up into droplets and water is dissociated partially into H^+ and OH ions supporting the molecular dispersion-based reactions of hydrolysis and hydrogenation of high-molecular fractions of organic energy carriers.

During cavitation treatment, the amount of free hydroxyl groups OH^- in water increases, and water acidity can increase to the ultimate value of pH = 14.0. During cavitation, collapsing cavitation bubbles yield important effects. These effects include: significant viscosity reduction; surface energy and tension change; inner energy decrease; the temperatures in the local collapse point rise to critical values of 374 K, at which the latent vaporization heat is practically absent; the water structure changes; intermolecular bond energy is released; water luminescence is observed; electric conductance increases, and chemical activity increases. The chemical compounds in water are activated, since there is practically no pure water, and in the natural state water is a solution of inert gases and different microelements. Reducing the heat of vaporization of cavitated water yields dramatic energy saving, since the heat of vaporization restores its prior value with subsequent condensation of the vapor obtained. The same pertains to reducing surface tension viscosity. This decreases the work expended on formation of vapor bubbles and their subsequent growth [5].

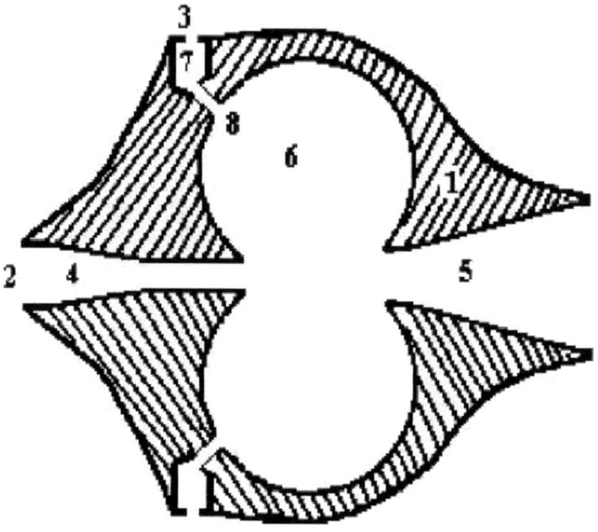

Fig. 3. Cavitation nozzle schematic

Experimental Results

The theoretical results obtained in [7] matched were experiments performed at the Laboratory for Gas and Flame Equipment (*FGUP VNIIAvtogenmash)*. The cavitation treatment of the mixture starts at 10-25 °C. The feeding pressure to the cavitator – 0-2 kgf/cm^2. (0 – in the case where only the cavitator is carried out the pump functions). For cavitation treatment we can use black oil, water, unconditioned hydrocarbons, refinery and coal industry wastes, tank washes, biomass and others.

Nozzles Used

Three samples of nozzles were used for testing. They were made to the above-described techniques (Fig. 4). The small-size nozzle – inner diameter of the toroidal chamber – 30 mm (external size is ≈3,5 cm in diameter). The medium-size nozzle – inner diameter of the toroidal chamber – 40 mm (external size is ≈5 cm in diameter). The large-size nozzle – inner diameter of the toroidal chamber – 50 mm (external size is ≈7 cm in diameter).

In Figure 4 the matchbox “КіпріДА” has dimensions of $5.5\text{cm} \times 3.5\text{cm}$ and is presented for understanding the physical size of nozzles. The physical sizes of nozzles are: 3-7 cm in diameter (external size). The nozzle sizes in Figure 4 are:

small-size nozzle – inner diameter of the toroidal chamber – 30 mm (external size is ≈3,5 cm in diameter).

medium-size nozzle – inner diameter of the toroidal chamber – 40 mm (external size is ≈5 cm in diameter).

large-size nozzle – inner diameter of the toroidal chamber – 50 mm (external size is ≈7 cm in diameter).

Fig. 4. External view of tested nozzles

Conduction of tests were made on a certified instrumentation and monitoring gas test bench. Compressed air was fed to the nozzles, and the atomized liquid was standard automobile diesel fuel.

The tests yielded the following results:

1. When scavenging the nozzles without feeding fuel it was found that, in the investigated air pressure range, there was rarefaction in the fuel channel. Its value increased with growth of initial air pressure at the nozzle inlet (Figs. 5 and 6). Figures 5 and 6 refer to the feed to the combustion chamber for different nozzle sizes.

2. Stable atomizing starts at pressures of about (0.3 – 0.5) kgf/cm^2.

3. During fuel atomizing two major fractions were distinguished, viz. the fine fraction with particle sizes of 5 μm to 10 μm, and a large fraction with particle sizes close to 100 μm. The fine fraction fills the central part of the atomized stream, practically without voids. The large fraction is distributed over the stream periphery and it is made up of droplets formed during disintegration of the film of the liquid phase. The liquid phase flows with low velocity over the walls of the nozzle outlet port and disintegrates under action of the inner part of the stream flowing out with high velocity.

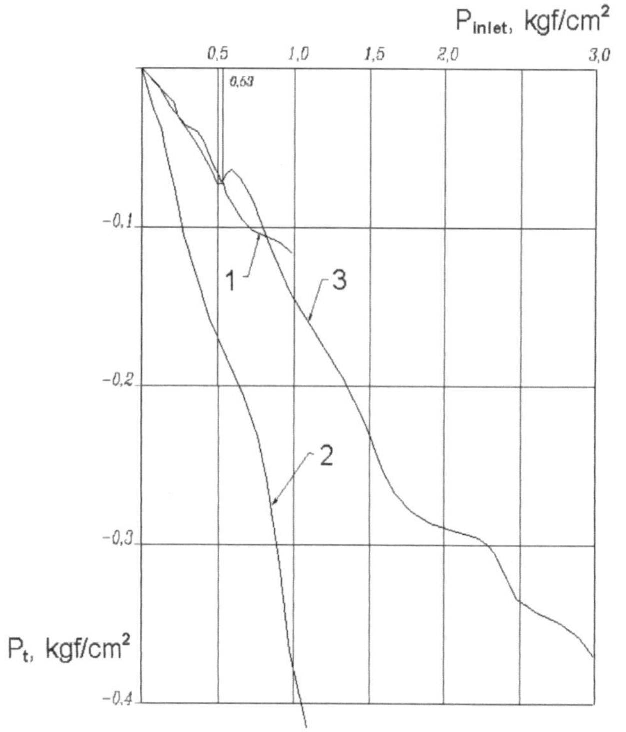

Fig. 5. Rarefaction in the fuel channel vs. nozzle inlet air pressure
1 – small-size nozzle; 2 – medium-size nozzle; 3 – large-size nozzle;
P_{inlet} – air pressure in inlet of the nozzles; P_t – pressure in toroidal vortex mixing chamber

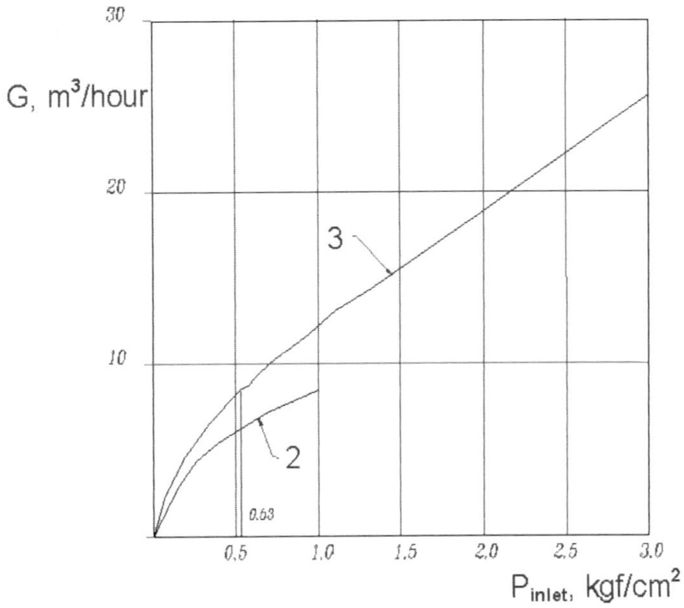

Fig. 6. Inlet air flow rate vs. Nozzle inlet air pressure: – small-size nozzle; 2 – medium-size nozzle; 3 – large-size nozzle; P_{inlet} – air pressure in inlet of the nozzles; G – air volumetric flow rate

To measure the size of the droplets we use the conductometric method and the laser method. The conductometric method is the better method to measure the size of the droplets in local zones, but the laser method is better for integral measurements.

Figure 7. Investigation of spraying dispersion using the laser method

The operating nozzles generated a loud noise, requiring using ear protection gear. The noise intensity and frequency changed depending on the air inlet pressure. The noise spectrum consisted mainly of high frequencies and frequencies close to ultrasound. No special measurements were made. However, even in absence of noise in the audible frequency range the rarefaction in the fuel channel fluctuated under definite conditions.

We are now developing artificial composite liquid fuel (ACLF) to demonstrate its effective combustion. ACLF is an emulsion or suspension comprising different combinations and content of brown coal, peat, bio mud, dung, tanker washout and other organic substances. These are water-black oil emulsions (WBOE); water-carbon-black oil suspensions (WCBOS); water-peat and water-bio mud mixtures, and others.

Refinement of Mixtures

Our laboratory studied the processes of obtaining and firing ACLF on the research set up developed by the Department for Nonconventional Power Technologies at IP-Mash NAS of Ukraine [7]. In addition, we present data on some experiments in fluid cavitation refinement of black oil to obtain distillate fractions.

The first run of experiments was conducted as follows: of the 100 % of a mixture of roughly 15 liters, 20 % accounted for black oil and 80 % was gasoline. Water was added in the amount of 25 % of that of black oil. Batch distillation refinement data was measured upon reaching a temperature of 80 °C, and a mixture density of 0.837175 kg/m^3.

The experimental data are given in Table 1.

Table 1. Experimental results of thermal draw refinement of the treated mixture: Distillate Fractions

	93 °C Initial Temperature
10 %	120 °C
20 %	137 °C
30 %	154 °C
40 %	175 °C
50 %	209 °C
60 %	251 °C
70 %	314 °C
80 %	335 °C
90 %	320 °C
	341 °C Final Temperature
Yield	94 %
Residue	5 %
Losses	1 %
Temperature in closed crucible – 14 °C	

The results indicate practically complete batch distillation refinements of black oil in a virgin petrol environment, yielding, mainly, the petrol draw and diesel fuel.

The second run of experiments (Table 2) was refinement of black oil in a diesel fuel environment with a 20 % addition of gasoline.

Table 2. Experimental results of draw refinement

Washout after refinement	Initial
I.c. 53 °C	I.c. 91 °C
10 % 79 °C	10 % 114 °C
20 % 95 °C	20 % 131 °C
30 % 108 °C	30 % 149 °C
40 % 131 °C	40 % 166 °C
50 % 187 °C	50 % 214 °C
60 % 266 °C	60 % 278 °C
70 % 330 °C	70 % 295 °C
82 % 358 °C	82 % 294 °C (initiation of coking)
F.c. 358 °C	$P^{20}{}_4$=0.7550 g/cm^3
Yield 90 %	Flash temperature in closed crucible – 18 °C
ρ=0.763 g/cm^3	Water – 0.018 %
Flash temperature in closed crucible – 22.5 °C	

In spite of refining black oil mainly in diesel fuel, more than 25 % of the black oil passed to the distillate draw from the initial 35 % of black oil.

ACLF was also studied with this same set up. ACLF containing bio-mud with a moisture content of 80 % and grade M100 black oil. The fuel obtained was combusted in a firing test bench, including the developed nozzles [7].

Combustion of Mixtures

The investigations demonstrate that the ACLF (20 % of bio mud and 80 % of M100 black oil) is close in fuel quality to M100 black oil. Certain characteristics of the composite fuel were identified when firing it (Fig. 7). The combustion heat of composite fuel was 9453 kcal/kg, being almost equal to the combustion heat of black oil M 100 (for black oil, 9530 kcal/kg). In getting into the combustion zone, water droplets (with a size up to 5 μm) explode. As a result, the fuel hydrocarbon fraction additionally disperses and sputters. In so doing, the combustion flare increases in volume by 11-12 % and fills the combustion chamber more compactly. The fraction of the heat accepted by the furnace increases by 7-10 %, and the hydrocarbon combustion degree reaches 99.5 %, and the boiler efficiency increases by 3-4 %. Stable combustion was obtained when using ACLF (50 % of bio mud and 50 % of M100 black oil) But, the combustion heat of this composite fuel is significantly less than the combustion heat of black oil M 100.

Fig. 8. Firing ACLF with a nozzle

Conclusions.

The studies conducted have shown that to create and effectively combust liquid fuel new fluid cavitation technologies are required. It is necessary to use nozzles-cavitators as atomizers during combustion. They do not require special conditions, i.e. heating. They even operate at subzero temperatures, which was confirmed experimentally. It was shown that, by achieving conditions of well-developed bubble cavitation, it is possible to improve atomizing quality and significantly change the chemical composition of combusted fuel mixtures.

Future Work

The studies presented in the paper and preliminary ACLF cost estimates allow for an optimistic view on the economic benefits of implementing the developments in the power industry. Tentative cost estimates for ACLF based on black oil, peat, sewage water dry residues, and other components are lower then black oil by 20%-25%. The cost estimates accounted for the following items: raw material, wages, electric power, and transport and overhead costs. Using the tank washes or old fuel oil, the ACLF cost can be reduced by 2-4 times compared to black oil.

The fluid cavitation technology of obtaining fine mixtures of different components is universal and can be used in different industries:

– preparing fuel emulsions for fuel economy enterprises;

– effective atomizing of medicinal preparations as a mist for husbandry and poultry farming;

– in the construction industry for making new materials, varnish-and-paint coatings, etc.;

– in the dairy and sugar-making industry, and in agricultural products processing;

– obtaining fine emulsions and suspensions for the food and cosmetics industry;

– of special interest is using the technology in the petroleum industry for reducing the viscosity of oil and oil products for their more efficient transportation.

References

1) H. Knickle**, "Oleg Kravchenko, krav@ipmach.kharkov.ua*, Iryna Suvorova, sig@ipmach.kharkov.ua*, Igor Baranov, bia.84@mail.ru*, *Podgorny Institute, Kharkov, Ukraine, **InEnergy Company, knickle@egr.uri.edu

2) I.G. Suvorova, O.V. Kravchenko, I.A. Baranov Mathematical and computer modeling of axisymmetric flows of an incompressible viscous fluid by the method of R-functions // Journal of Mathematical Sciences. Springer US, 2012. – Vol. 184. – No. 2. – pp. 165-180.

3) V. L. Rvachev, I. G. Suvorova, T. I. Sheiko. Numerical realization of the problem of the flow of a viscous fluid by the R-function method in the "field-3" system // Journal of Mathematical Sciences. Springer US, 1993. – Vol. 65. – Issue 33. – pp. 1905-1908.

4) V. L. Rvachev, L. I. Kurpa. Special-purpose interactive programming shell for the "pole" system // Cybernetics and Systems Analysis. Springer US, 1995. – Vol. 31. – Issue 1. – pp. 131-137.

5) Patent Application № 99,111,902/28 RF, IPC^7 B63B59/08, B08B3/02. Device for treatment of surfaces immersed in a liquid / A.I. Kharlamov (Russia). – Filing date 15.06.1999.

6) Utility model patent № 18,922, Ukraine, IPC^7 B01F 3/08, B 63 B59/00/ Mixer nozzle / I.G. Suvorova (Ukraine), O.V. Kravchenko (Ukraine); A.N. Podgorny Institute for Mechanical Engineering Problems NAS of Ukraine. – № U 200,606,857; filing date 19.06.2006; published 15.11.06, Bull. № 11. – 4 p.

7) Patent Application № 200,510,753 (UA), IPC^7 B01F7/00, C10G7/06. Method for refinement of black oil and a rotor-cavitation disperser therefor /Miroshnichenko I.I., Suvorova I.G., Matsevity Yu.M., Kravchenko O.V., Tarelin A.O. (Ukraine). – Filing date 14.11.05.

Advanced Aromatics Recovery Technology with AED-BTX Process

CIPUTRA JAP
JEFFREY CHIU
KUANG WU
AMT INTERNATIONAL, INC.
PLANO, TX 75024
U.S.A.

SUNG-CHUL YOO
LG CHEM, LTD.
YEOSU, KOREA

TZONG-BIN LIN
JYH-HAUR HWANG
CPC CORPORATION
TAIPEI, TAIWAN

PREPARED FOR PRESENTATION AT
THE 2015 AICHE SPRING MEETING
IN AUSTIN, TX, APRIL 28TH, 2015

Unpublished

ABSTRACT:

For over forty years, the petrochemical industry has used Liquid-Liquid Extraction (LLE) for aromatics recovery. Extractive Distillation (ED) was developed in the 1950's, but due to a lack of expertise in two–liquid phase distillation and ED process simulation/modeling, the implementation of ED was not widely used until recent years.

AMT International, Inc. (U.S.A.) and CPC Corporation (Taiwan) have developed a novel Advanced Extractive Distillation Process, referred to as the *"AED-BTX Process"*, for effective aromatics recovery with substantial energy savings. One of the key features of the Process is the use of the original Sulfolane solvent as the extractive distillation solvent to achieve high selectivity without any special solvent or additives.

The AED-BTX Process has been successfully applied in LG Chem's BTX Extraction unit in Yeosu, Korea in 2013. Specifically, the Process has achieved and exceeded all process requirements with over 35% energy savings when compared to the original LLE Extractive Stripper. Furthermore, by reusing the existing Sulfolane solvent and most of the LLE equipment, the application of the AED-BTX Process in LG Chem's BTX Extraction unit has resulted in a payback period of less than six (6) months.

This Paper is being co-authored by AMT, LG Chem, and CPC to illustrate the Process, plant modifications, and improvements made through the conversion of LG Chem's LLE Process to ED Process.

KEYWORDS: *aromatics, BTX, sulfolane, liquid-liquid extraction, extractive distillation*

Tel: 972-378-0700, Fax: 972-378-9400, E-Mail: information@amtintl.com

398353 Cost Effective Control of Red Oil in Caustic Tower Systems

Abstract:

Cost Effective Control of Red Oil in Caustic Tower Systems

Authors: **Kyle David Mankin** Global Marketing Manager- Petrochemicals

P. N Ramaswamy, Global Technical Director

Saurabh Shende, Technical Manager – Petrochemicals

Abstract:

In the modern era, the economics for producing olefins have become more of a challenge. To maximize profit, ethylene crackers are being asked to process "opportunity naphtha's" or other liquid feeds which are available at advantaged pricing. Often, these feeds contain elevated levels of sulfur which lead to higher production of H_2S in the furnaces. In some regions, producers are also turning to new technology for producing olefins from methanol. These drastic design changes are creating challenges in the operations of normal caustic tower systems, which are critical for the removal of contaminants like H_2S and CO_2 which are generated in furnaces.

One of the most frequent causes of mal-operation in a caustic tower system is the excessive formation of red oil polymer. Red oil polymer is generated due to Aldol condensation of carbonyl compounds including a mix of various ketones and aldehydes. Conventional treatments for control of red oil formation have focused on the reduction of aldehydes to alcohols or dispersing the polymers after they are formed. Dorf Ketal has developed a patented new technology that inhibits polymer formation while also scavenging polymer molecules already in existence. With this new approach, our chemistry is effective even when applied in sub-molar ratios, thus reducing the treatment cost without sacrificing efficiency.

This paper will focus on the sources and mechanisms of fouling in a caustic tower system and will show examples of the effectiveness of our innovative new treatment program including a review of data and experience from a customer operating a 900 KTA Ethylene facility.

401801 Separation of Propylene/Propane in Facilitated Transport Membrane Containing Silver Nanoparticle

Abstract:

Membrane-based gas separation enables a chemical process to be low-energy consuming and environmentally friendly while C_3 splitter requires high energy consumption in petrochemical processes. This study is aimed to separate propylene and propane using membrane process instead of NCC(Naphtha Cracking Center) C_3 splitter. In this study, high performance facilitated transport membrane (FTM) is introduced for propylene/propane separation. We fabricated FTM composite membrane module using PVP/$AgBF_4$/TCNQ composite on top of hollow fiber membrane. Pure gases and mixed gas permeation test were performed. We developed simulation program predicting the membrane separation properties under operation conditions. Separation properties of FTM module for propylene and propane were obtained from the simulation program based on the pure gas permeation data. Based on these results, it is predicted that an one-stage membrane process provides 99.5% of propylene at permeate side from a binary gas mixture of 95 vol% C_3H_6 / 5 vol% C_3H_8 supplied as a feed gas.

400705 A New Era in on-Purpose Butadiene Technology

Abstract:

Low cost natural gas production has lead to an abundance of associated natural gas liquids (C2-C4) resulting in displacement of naphtha as feed for steam crackers. This shift in feedstock results in decreased byproduct butadiene production, causing production to be unable to keep pace with rising demand. This 'butadiene gap' can be closed with new technologies to produce on-purpose butadiene. In 2014, TPC Group Inc. and UOP LLC, a Honeywell Company, entered into an agreement to develop and license the most efficient and low-cost method for on-purpose butadiene production, leveraging UOP's expertise in licensing and technology. An improved commercial process is now being offered that builds on the proprietary TPC Group OXO-D™ technology, which has been successfully operated for more than 40 years, commercially converting butene to butadiene for the production of synthetic rubber. This presentation will provide a description of this new process and highlight differences and the benefits compared to the previously commercialized OXO-D technology. In addition, key development efforts and demonstrated operating performance will be highlighted.

Value Cracking – Make Liquid Crackers competitive

Heinz Zimmermann
Linde Engineering
27 th EPC Conference, April 27 – 30, 2015, Austin / Tx

THE LINDE GROUP
Linde

Steam cracker economics summary: Cash costs vs global capacity - Naphtha Cracker set price

THE LINDE GROUP
Linde

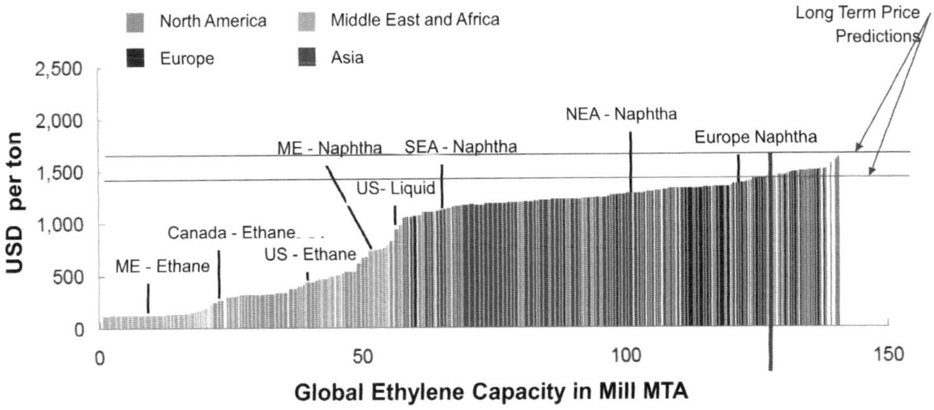

Source: McKinsey CMAI, Platt's

Analysis: Naphtha / NGL crackers need improvement in cash costs

THE LINDE GROUP Linde

Measures are:

1. Lower Naphtha / LPG prices
2. Flexibility in feedstock's
3. Integration into refinery
4. Produce more valuable products

No: 3 Value Cracking Linde Engineering

Mild Cracking to produce more Propylene and Butadiene

THE LINDE GROUP Linde

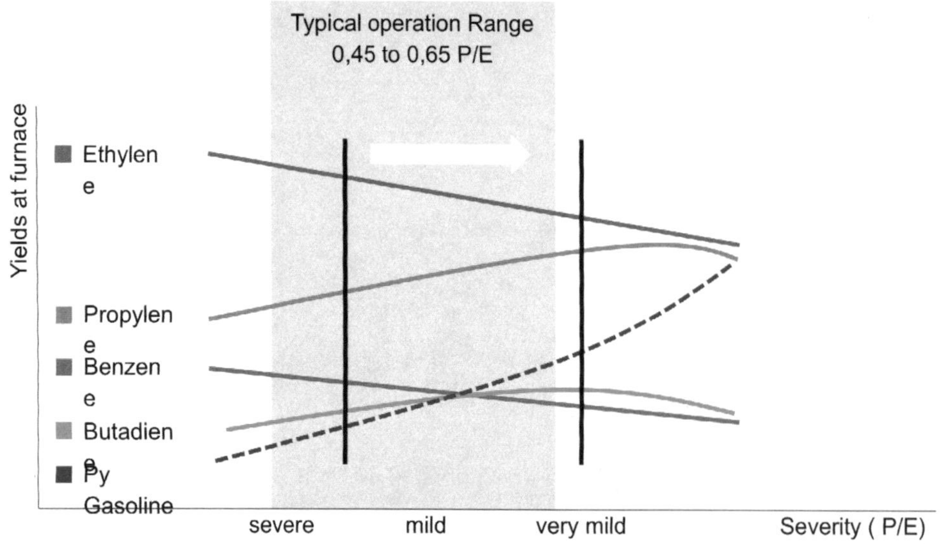

No: 4 Value Cracking Linde Engineering

Recycle Management is key element in Mild Cracking

THE LINDE GROUP
Linde

— **Mild Cracking: Increased Recycle Streams**

Decreasing Concentrations of Value Products (extraction)

➡ **Smart Recycle Management**

Method developed by Linde based on Lab cracking results with many individual components

No: 5 Value Cracking Linde Engineering

Butane / Butene Cracking Pilot Plant Results

THE LINDE GROUP
Linde

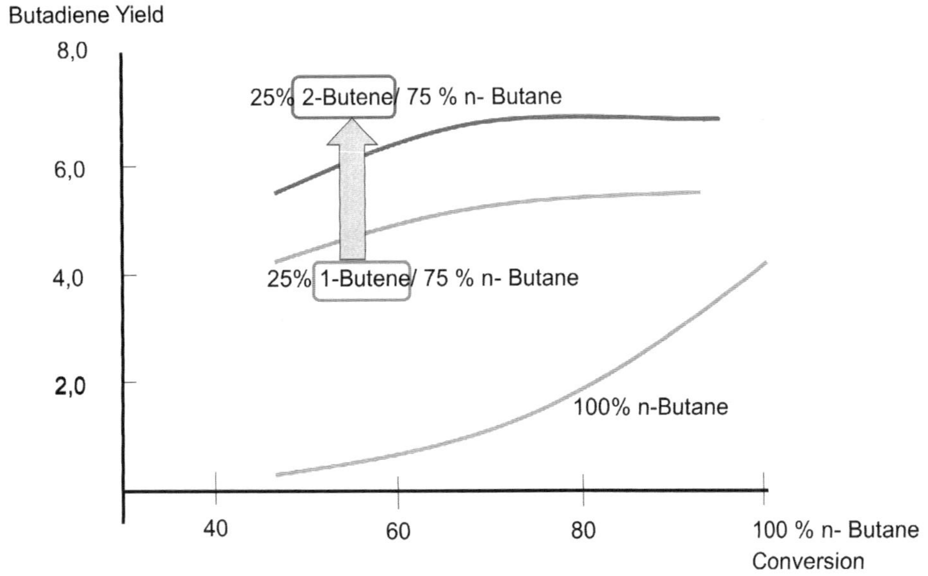

No: 6 Value Cracking Linde Engineering

Pilot Plant Cracking Results

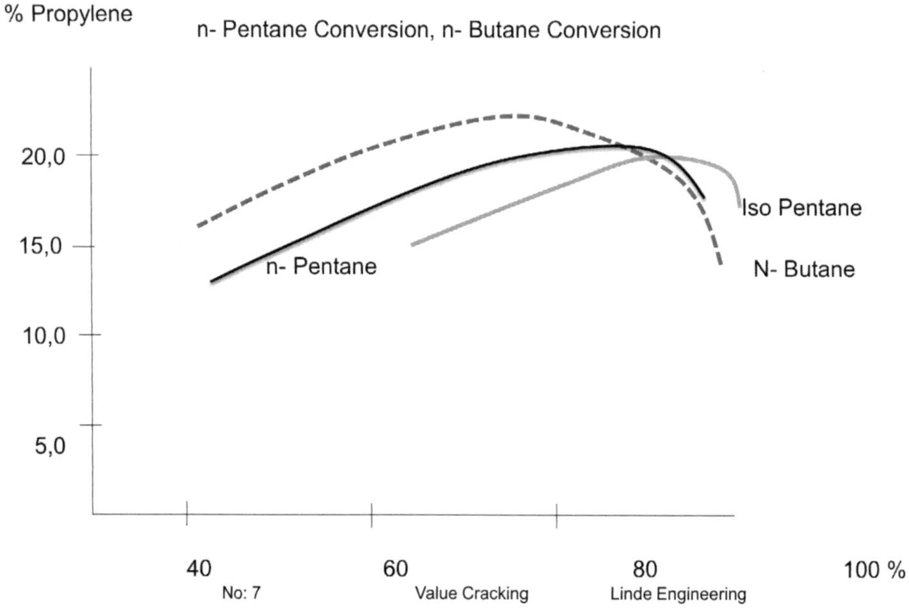

Principles derived from Pilot Plant results

— **Maximum Butadiene : and** — **Alpha Olefins converted to internal Olefins extreme mild cracking**

— **Maximum Propylene:** — **N- C4s Mild Cracking,**

Iso-C4s, C5s Severe Cracking

Severe Cracking of some recycles is important to increase the concentration value products

Principle of Recycle Management
Separate, treat and recycle to dedicated furnaces

THE LINDE GROUP
Linde

e.g. C4 Fraction: Butadiene Yield maximization

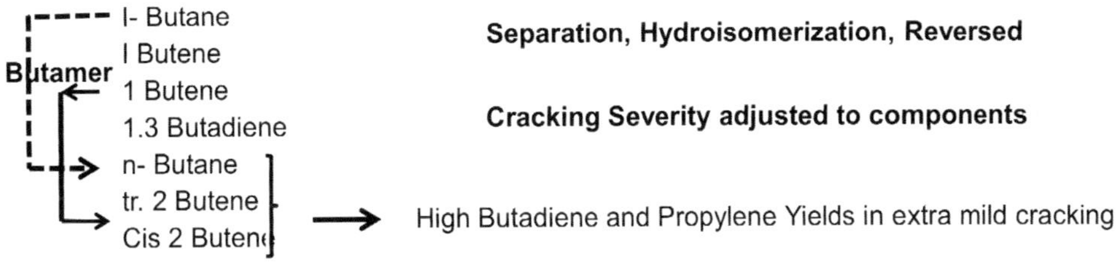

Separation, Hydroisomerization, Reversed

Cracking Severity adjusted to components

No: 9 Value Cracking Linde Engineering

New Development : Individual Routing of Components
e.g. : C4 Fraction to increase High Value Products

THE LINDE GROUP
Linde

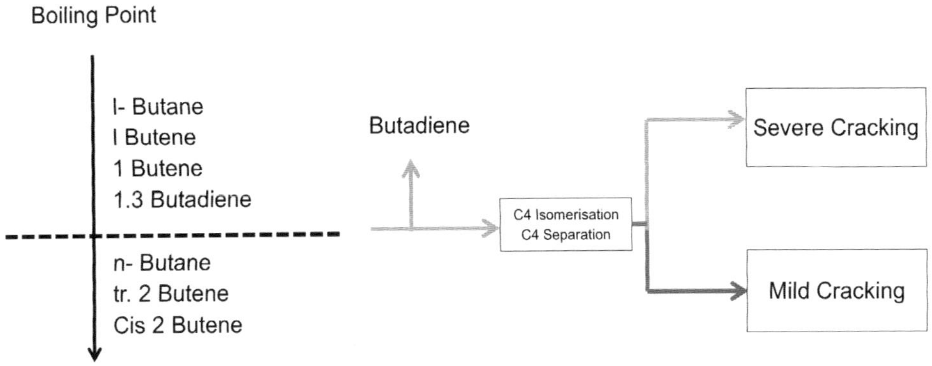

No: 10 Value Cracking Linde Engineering

Naphtha Cracking : All Cases

THE LINDE GROUP

Linde

Feedstock: Full Range Naphtha

2.697 mill MTA

No: 11 Value Cracking Linde Engineering

Base Case: Naphthacracking

THE LINDE GROUP

Linde

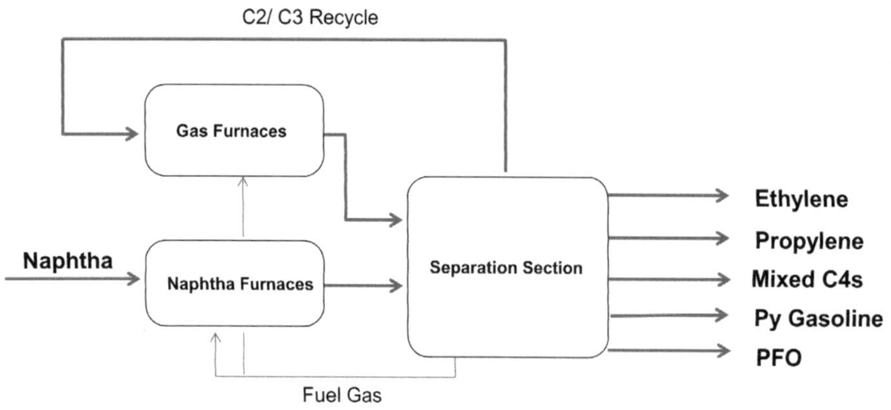

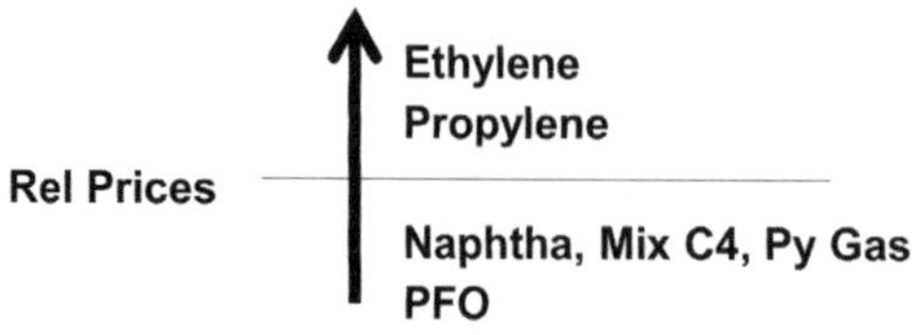

No: 12 Value Cracking Linde Engineering 12

Step 1: Add Butadiene Extraction
Step 2: Add Benzene Extraction

THE LINDE GROUP Linde

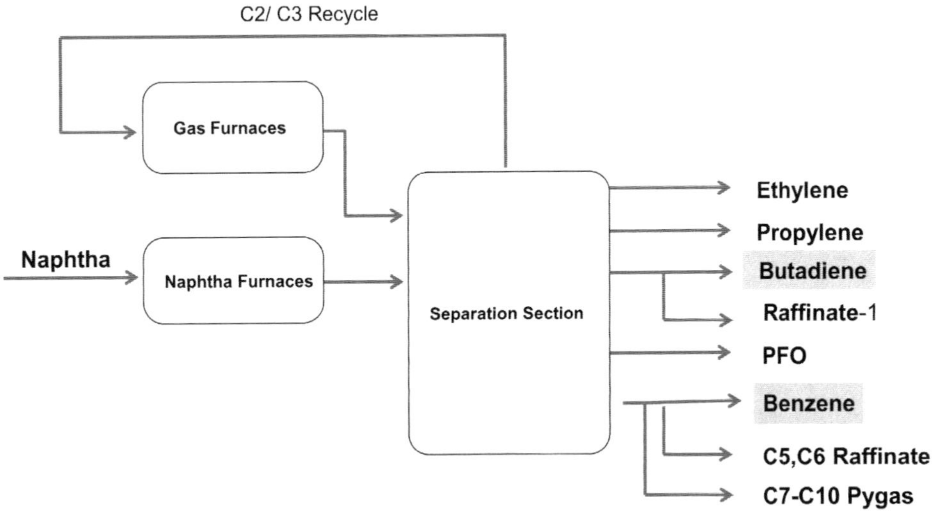

No: 13 Value Cracking Linde Engineering

Step 3: Recycle Raffinate -1
Step 4: Recycle C5, C6

THE LINDE GROUP Linde

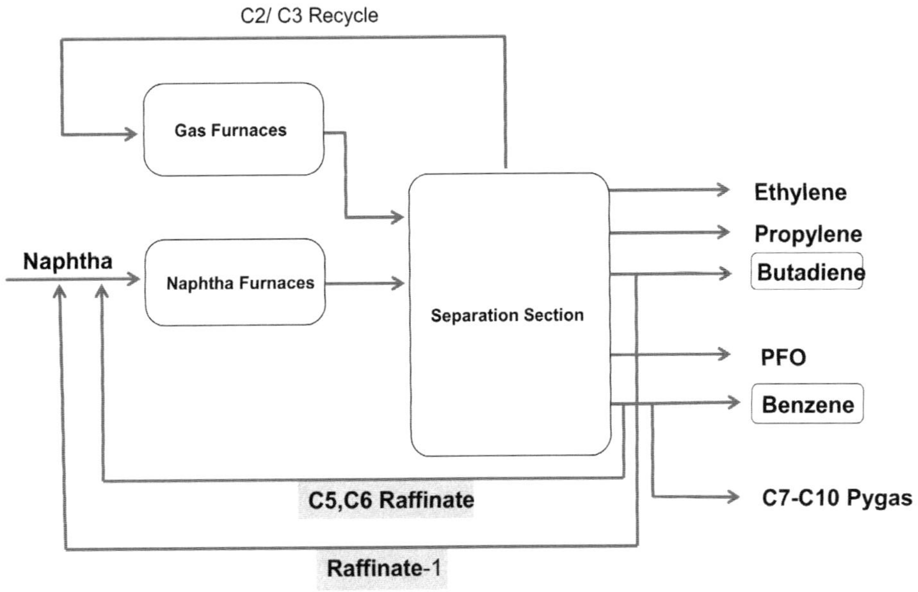

No: 14 Value Cracking Linde Engineering

Step 5: Extract BTX and recycle C5, C6-C8
Step 6: Recycle C5, C6

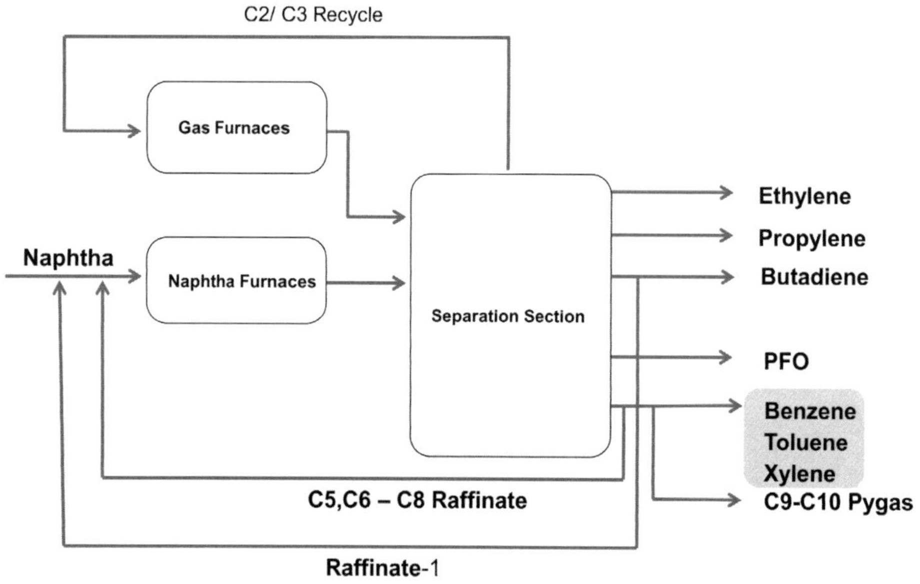

No: 15 Value Cracking Linde Engineering

Step 7: Value Cracking Scheme , C4 Iso and Separation, Individual recycle severity (Patents filed)

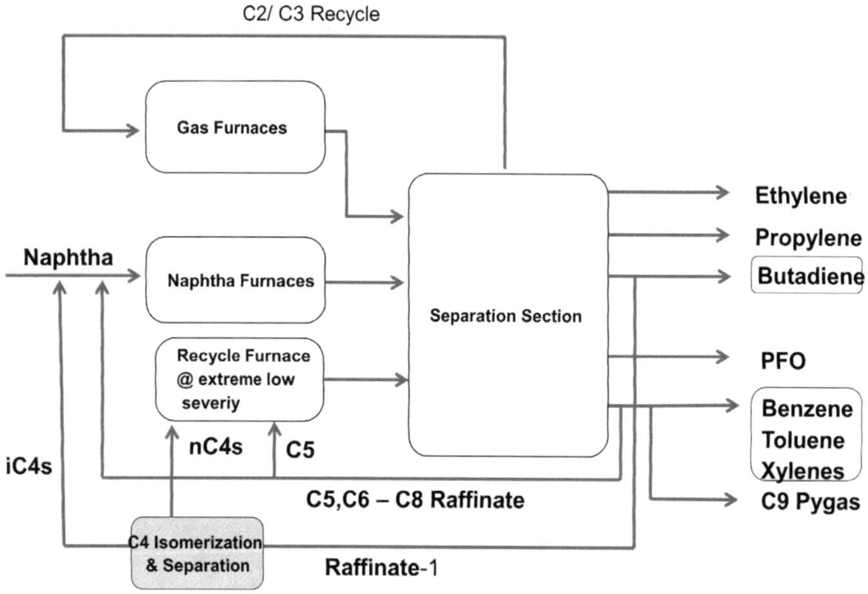

No: 16 Value Cracking Linde Engineering

Step 8: Additional C5 Separation

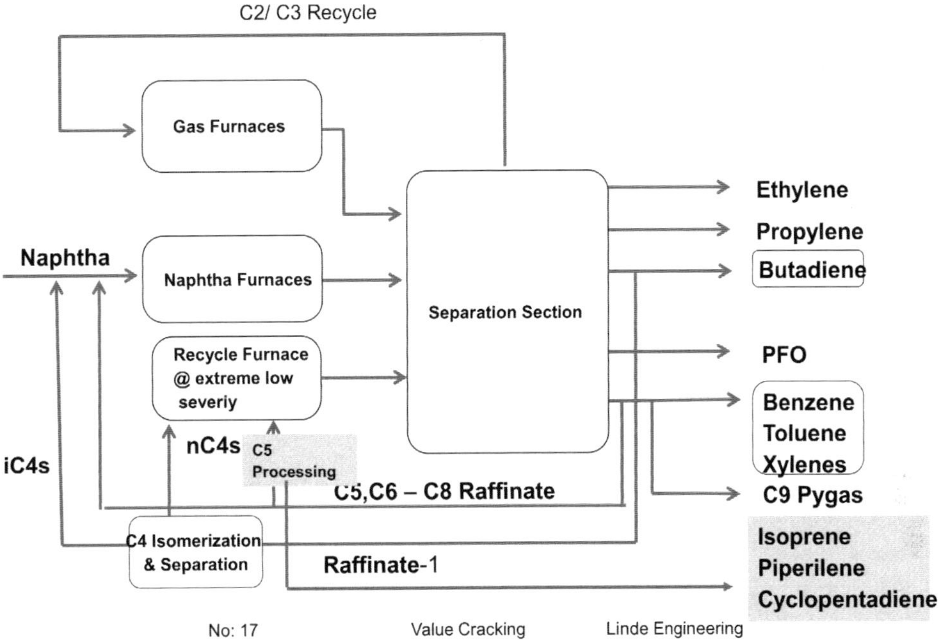

No: 17 Value Cracking Linde Engineering

C5 Processing (GTC Process)

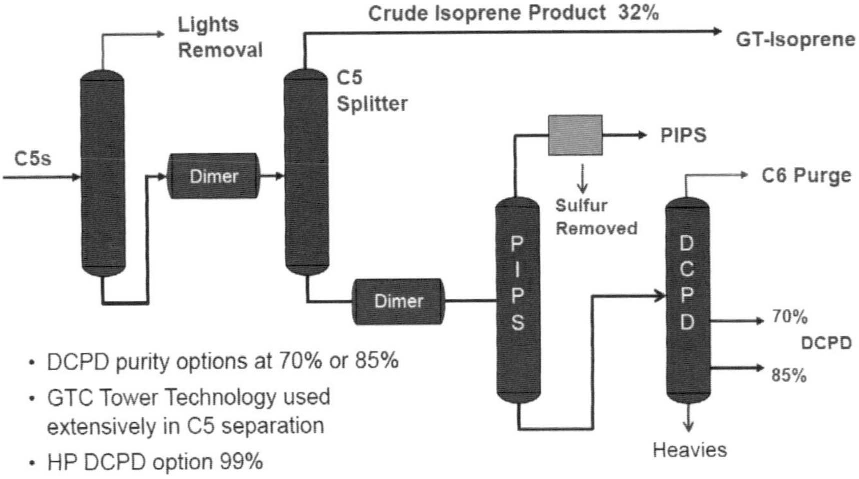

- DCPD purity options at 70% or 85%
- GTC Tower Technology used extensively in C5 separation
- HP DCPD option 99%

No: 18 Value Cracking Linde Engineering

Step 9: Additional Reversed Butamer

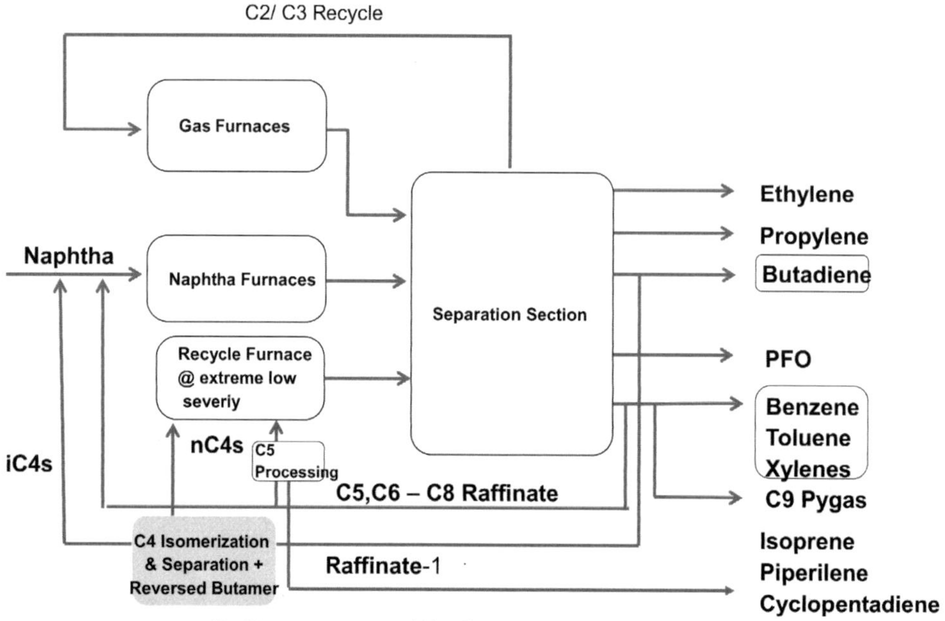

No: 19 Value Cracking Linde Engineering

Naphtha Cracking: Products from 2.697 mill MTA

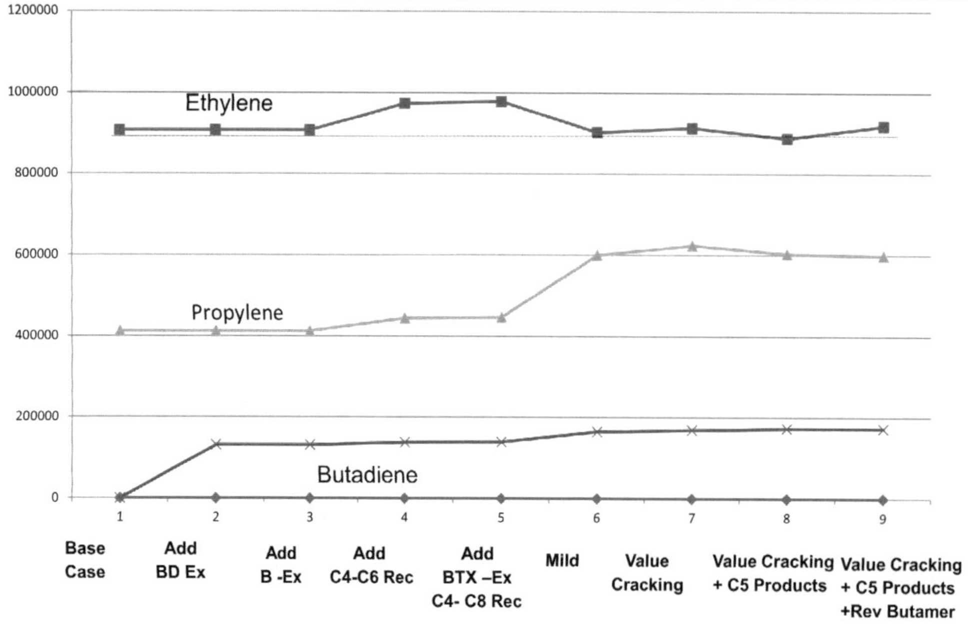

No: 20 Value Cracking Linde Engineering

Naphtha Cracking: Products from 2.697 mill MTA

Case 8,9: 26 359MTA Isoprene, 19622 Piperilene, 35 885 CP

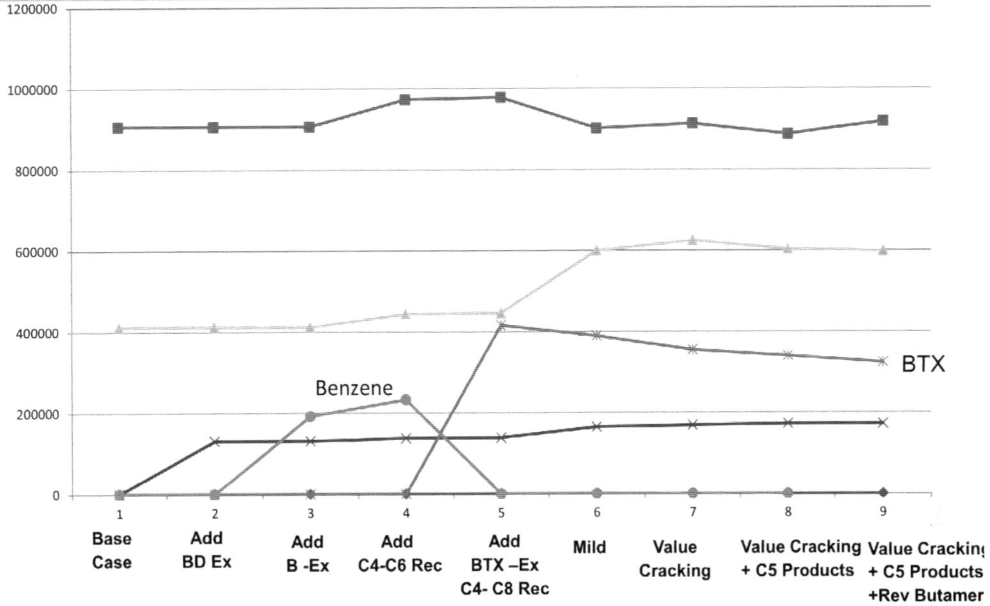

No: 21 Value Cracking Linde Engineering

Prices used for evaluation

	High Oil Price 91 USD / bbl	Low Oil Price 60 USD/bbl
Naphtha	800	524
Ethylene	1350	842
Propylene	1450	808
Butdiene	2000	786
BTX	1100	660
Benzene	1200	720
Pygas	800	542
C4- Fraction	880	576
Isoprene	2500	1500
1,3 PD	1400	840
1,3 Cyclo PD	1400	840
PFO	600	400

No: 22 Value Cracking Linde Engineering

Economics of options vs base case: Naphtha @ 91 USD /bbl

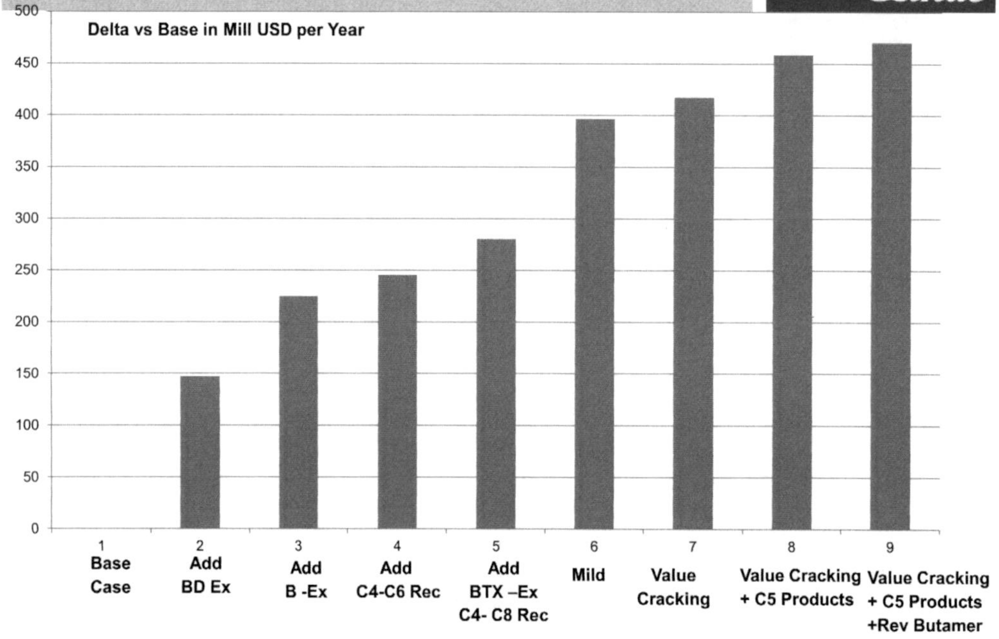

No: 23 Value Cracking Linde Engineering

Economics of options vs base case: Naphtha @ 60 USD /bbl

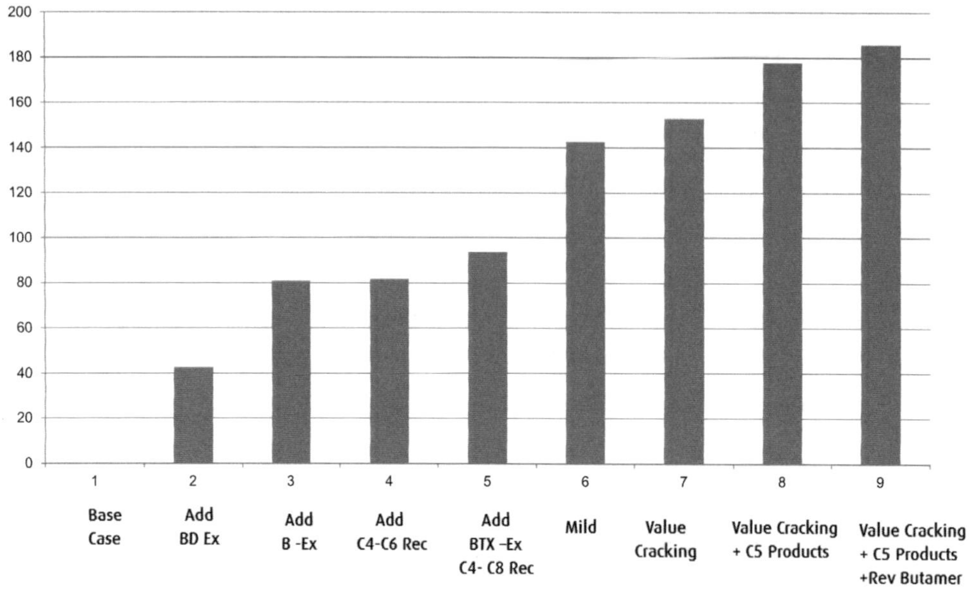

No: 24 Value Cracking Linde Engineering

Naphtha Cracker Economics @ 91 USD/ bbl Crude

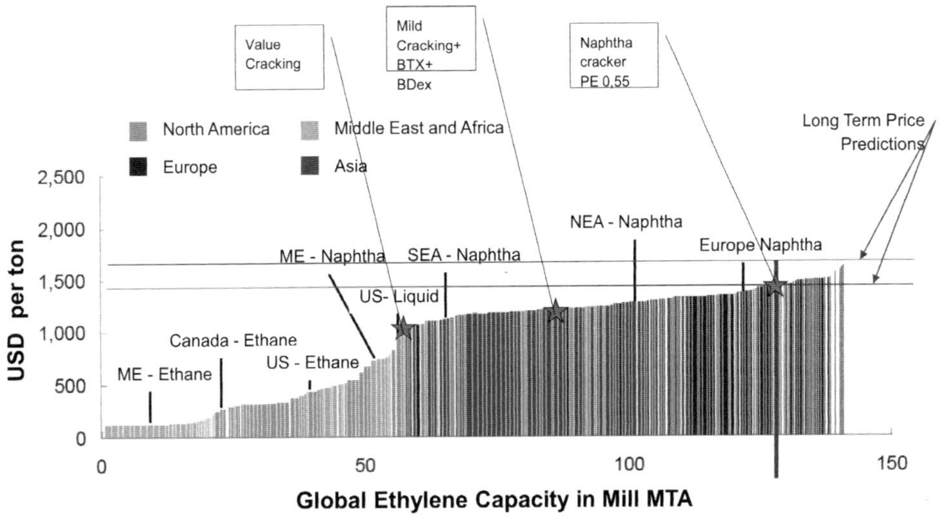

Source: McKinsey CMAI, Platt's

No: 25 Value Cracking Linde Engineering

Naphtha Cracker Economics @ 60 USD /bbl Crude

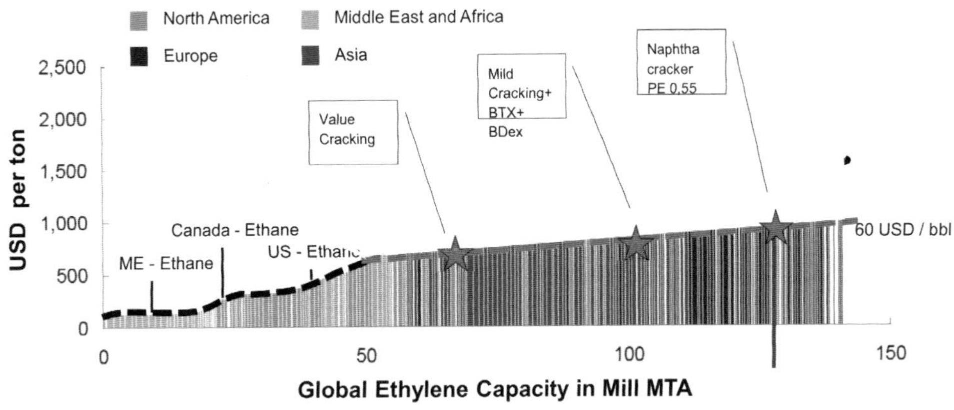

Source: McKinsey CMAI, Platt's, Linde Analysis, HSBC

No: 26 Value Cracking Linde Engineering

Summary

- Steam cracking of liquid feeds consequently targeting for Hi value products
- New patented scheme to improve cracker economics significantly
- The scheme has advantages for Naphtha cracker in high and low crude price scenarios
- Scheme can be applied to existing crackers / tailored solution
- Details available under confidentiality agreement

No. 27 Value Cracking Linde Engineering

Thanks for your attention.

About Popcorn Polymer

The issue at a glance

Bernd Heida, BASF SE

BASF
The Chemical Company

Content

BASF
The Chemical Company

- What is popcorn polymer
- Reasons for popcorn formation
- Prevention & special items
- KPIs

2

Where is popcorn polymer?

BASF The Chemical Company

■Popcorn polymer

- Is an issue in butadiene plants
- Is frequently an issue in butadiene processing plants
- Is incidentally an issue in steam crackers
- Is seldom an issue in storage and transportation of butadiene

3

What is popcorn polymer?

BASF The Chemical Company

■Popcorn polymer

- is a crosslinked polybutadiene
- reaction requires high butadiene concentration to start
- is a hazard to the plant, depending on its location in the plant and the degree of crosslinking
- may be a hazard when removed from the plant
- has an appearance resembling popcorn
- is of white color when active, getting yellow when deactivated
- may contain liquid butadiene peroxides

4

What type of popcorn polymer?

■Appearence

- Popcorn seeds and low crosslinked popcorn show crystal clear appearence and are elastic similar to car tyres.
- Popcorn with a high degree of crosslinking is opaque, rigid, of very irregular shape and porous.
- In case of popcorn formation close to a major source of oxygen, butadiene peroxides may be formed in parallel, being trapped in the pores of highly crosslinked popcorn.

5

Popcorn damage on condensers

Content

BASF
The Chemical Company

- What is popcorn polymer
- Reasons for popcorn formation
- Prevention & special items
- KPIs

7

What starts popcorn formation?

BASF
The Chemical Company

■Free radical starter

- Oxygen is the most critical molecule. It may generate peroxides or other reactive molecules containing oxygen who finally start the polymerization.
- Haematite and blank shining carbon steel surface are considered to promote formation of highly crosslinked polymerization
- Deactivated popcorn is never 100% deactivated!

8

Popcorn formation through oxygen inclusion

BASF
The Chemical Company

Level indicator:
Popcorn destroys gasket →damage on membrane

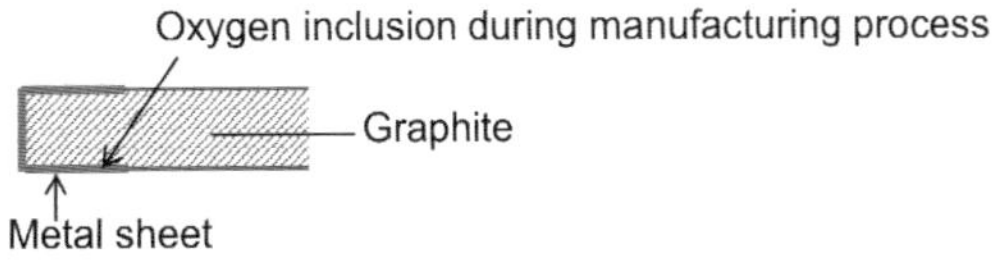

9

Content

BASF
The Chemical Company

- What is popcorn polymer
- Reasons for popcorn formation
- Prevention & special items
- KPIs

10

Pickling and passivation

Improper P&P leads to popcorn formation

- Incomplete removal of Haematite from metal surface
- Remaining areas of blank shining metal surface
- Insufficient flushing
  - Remaining pieces of haematite
  - Remaining reactive chemicals

11

Popcorn prevention

The role of sodium nitrite

- It is an oxygen scavenger
  - **In BDX, aqueous solution used to be the inhibitor in distillation section.**
  - **Is frequently used in BDX as anti foulant additive in the solvent**
- It is a passivator (build up of a protective layer)
  - **Very efficient in aqueous solution, but causes problems in waste water treatment plants**
  - **Today often used in combination with chemical mixtures who build up magnetite**
  - **Alternatives: other nitrites, phosphites, amines, products from service companies**
- There may be safety issues
  - **Used solutions may contain sodium nitrate**
  - **Depending on composition, concentration and temperature, sodium nitrite and sodium nitrate may decompose / cause explosion.**

12

Popcorn prevention

BASF
The Chemical Company

The role of TBC

•It is an oxygen and free radicals scavenger

- In BDX, it is widely used as an oxygen and free radicals scavenger in popcorn risk areas
- It is an agressive chemical, may remove protective layers / passivation layers from the carbon steel surface → combined application with DEHA may be useful

•It is a solid, it's solubility in HC's is low

- Precipitation in the bottom of heavies removal columns happens, leading to plugging of the bottom withdraw line

•Alternatives: DEHA, various products from service companies

13

Popcorn prevention

BASF
The Chemical Company

Keep the plant tight

•Ensure all flanges to be as tight as possible

- Soap test is not a guarantee of sufficient tightness
- Manholes may need flange facing
- Constant flushing with nitrogen of critical flanges

•Avoid oxygen ingress from sources like

- External feedstock
- Maintenance (repairs / changes of equipment)

•Avoid locations of uninhibited liquid BD

14

Popcorn prevention

BASF
The Chemical Company

The role of uninhibited butadiene

•Without inhibitor, popcorn formation starts much easier and propagates much faster.

•Areas of uninhibited liquid HC's are cold spots

- **Condensers**
- **Manholes**
- **Flows to columns being colder than the temperature in the column**
- **Zones of stagnant butadiene vapor phase**

15

Popcorn prevention

BASF
The Chemical Company

Butadiene at low concentration – be aware of!

•Few cases of popcorn formation at low butadiene level are reported

•Popcorn seeds flushed to areas of low butadiene concentration will continue to grow

•Butadiene plus high boiling components (e.g. solvent)

- **Low butadiene content at high temperatures may cause high butadiene content at low temperatures. Such may happen for example at the end of a connected line with no flow**

16

Popcorn prevention

BASF
The Chemical Company

Avoid formation of peroxides

•Peroxides are an excellent starter for popcorn polymerization

•Oxygen ingress causes formation of butadiene peroxides

•Mostly they appear as an oily liquid, barely soluble in HC.

•Popcorn soaked with peroxides may explode at mechanical shocks → prior destruction to be covered with caustic solution or with water

17

Operational negligence

BASF
The Chemical Company

Not to forget:

- hydrocarbons out = oxygen in
  - **Diffusion velocity = f(Δc)**
- Cooling water = oxygen carrier
- Accumulation of trace components / byproducts: There are always starters included
- Incomplete removal of popcorn despite deactivation
  - **Popcorn has pores and micro pores: deactivation efficiency never can reach 100 %**

18

Content

BASF
The Chemical Company

- What is popcorn polymer
- Reasons for popcorn formation
- Prevention & special items
- KPIs

19

KPI's

BASF
The Chemical Company

Operational excellence

•Avoid oxygen in the feedstock or remove it by pretreatment

•Cooling water entrainment

- **Provide a good and reliable monitoring system**
- **Remove completely or replace leaking tubes.**

•Avoid accumulation of unknown chemicals mixes in the process (external ingress, internal accumulation)

•Provide procedures and schedule for popcorn prevention

- **Flashing of dead ends, DP observation, pump filter check, equipment checks in the field, analysis program, etc, etc.**

20

KPI's

Maintenance excellence

•Develop a plan how to improve the tightness of the plant

- **Single measures, schedule for periodic actions**

•At all maintenance activities, strictly avoid oxygen ingress ito the plant

•Only use spare parts in perfect condition

•Do not repair bulged equipment (e.g. heat exchangers)

•Passivate all spare parts or new equipment prior start of operation

21

KPI's

Equipment quality

- Tightness (to avoid oxygen ingress)
- Cavities potentially accessible for butadiene (e.g. on valves)
- Possibility of oxygen inclusion (e.g. gaskets)
- General quality of the applied materials

22

Thank you!

Any more questions?

23

Development of highly selective and stable DTP catalyst for propylene production from methanol

Dr. Kazunori Honda[1]*, *Dr. Nobuyasu Chikamatsu*[1], *Mitsuo Morita*[1] *and Shuji Obayashi*[2],
[1] *Technology Development Center, JGC Corporation, Yokohama, Japan*
[2] *Mizushima R&D Center, Mitsubishi Chemical Corporation, Kurashiki, Japan*
*E-mail: honda.kazunori@jgc.com

Introduction

The global demand for propylene is forecast to continue growing strongly, and the supply/demand balance for propylene is expected to remain tight. Propylene is produced mainly from petroleum-based raw materials through naphtha crackers or Fluidized Catalytic Cracking (FCC) used in petroleum refining, but is a by-product in either case, and it is difficult to boost production of propylene alone with these manufacturing methods. In addition, diversification of propylene feedstock is needed as crude oil prices have soared. With this background, attention has become focused on the diversification of feedstock to include carbon sources other than petroleum and on the development of processes whose purpose is directly oriented toward propylene production. Recently, on-purpose ethylene/propylene production technologies from DME (dimethyl ether) /methanol, which are produced from low cost feedstock in comparison with oil such as coal, are being introduced in China. On-purpose propylene production project using competitive methane derived from shale gas is expected in North America.

JGC Corporation and Mitsubishi Chemical Corporation have jointly developed the DTP (Dominant Technology for Propylene Production) process in which propylene with high yield is produced by using DME or methanol. The basic process design is derived from the data gathered at a demonstration plant constructed at Mitsubishi Chemical's Mizushima Plant. Continuous operation had been successfully conducted and the accumulated operation time was over 12,000 hours. The DTP process is ready for commercialization.

For the commercial condition, low propylene selectivity for the initial period of reaction due to the high activity and insufficient catalyst life by low hydrothermal stability are main issues to improve catalyst performance. Therefore, we continue to improve catalyst performance and have developed a promising catalyst. The concept of the catalyst design is "moderate active" but "stable". The new catalyst has high and stable selectivity to propylene from start of run to end of run and longer catalyst life. In this paper, features of new catalyst will be provided.

Process Description

Process flow

The simplified flow of DTP process is shown in **Fig. 1**. First, feedstock methanol is converted to DME where a state of equilibrium is established among the DME/water/methanol components at the methanol dehydration reactor. Then, the recycled hydrocarbon is added to the DME stream, and the mixture is fed into the DTP reactor to produce propylene. The methanol/DME conversion reaches 99.99% in the reactor. As the methanol conversion is extremely high, waste water can be treated without complex treatment. The products from DTP reactor are separated and propylene can be obtained. Among the by-products, hydrocarbons especially C_4-C_6 olefins are recycled to be converted to propylene. Optionally, C_4-C_6 olefins from another source can be added as feedstock.

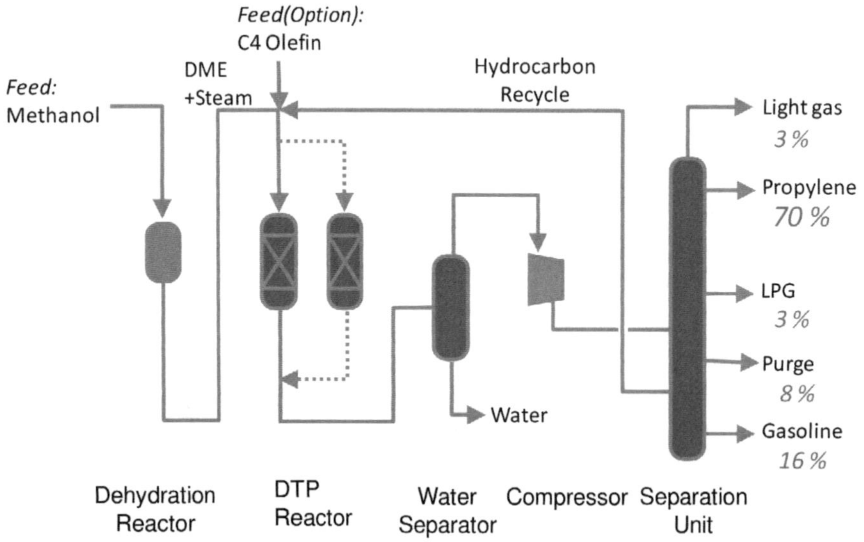

Fig. 1 Process flow diagram of DTP unit

DTP Reactor

Simple fixed bed adiabatic reactors are applied to the DTP reactors, for which the operation is easy. DTP catalyst is a proprietary MFI type zeolite. During the reaction, coke is accumulated on the catalyst. Before any significant fall in the reaction efficiency, the in-operation DTP reactor is exchanged. The catalyst in the reactor is regenerated by burning the coke with diluted air. For continuous operation, one or more reactors are in operation while the other reactor is undergoing regeneration or in a stand-by mode.

Experimental

Preparation of DTP catalyst

Catalysts based on metal modified ZSM-5 zeolites were used in this work[1)]. Two types of metal modified ZSM-5 with different metals were synthesized by the hydrothermal treatment of a suspension of the starting materials in an autoclave at 413 K for 12 h. The resulting zeolites were extruded with binder materials, crushed and sieved to collect fractions of 250-500 μm before use. In the sample, names DTP-1 and DTP-2 denote the current catalyst and the newly-developed catalyst, respectively.

Steam dealumination

It is known that aluminium cation is removed from the zeolite framework by the exposure to steam[2)] and then the zeolite as the acid catalyst deactivates irreversibly. To investigate the hydrothermal stability of the zeolite based catalyst, the dealumination from the zeolite framework was accelerated by the exposure to steam with severe condition. Steam dealumination of zeolites was conducted at a condition of 0.56 MPa (partial pressure of steam was 0.53 MPa), 833 K for 50-200 h.

Catalytic performance test

A two-stage fixed-bed continuous flow reactor connected in series was mainly used for reaction tests in order to examine catalytic performance under simulated commercial conditions. Feed methanol, steam and nitrogen as internal standard were supplied to the first reactor filled with gamma alumina at 573 K with a WHSV of methanol in the first reactor at 0.37 h^{-1}. Part of methanol was converted to DME and steam within the first reactor, methanol conversion was stably kept at almost 70% through the experiments. In addition to the effluent from the first reactor, isobutene as a simulant recycle component was fed to the top of second reactor at 753 K. The molar ratio of feed methanol, isobutene, steam and nitrogen were 1.00:0.24:1.26:0.22 and a WHSV of methanol in the second reactor was 0.69 h^{-1}. Methanol, DME and isobutene were converted to lower olefins such as ethylene, propylene by contact with the DTP catalyst filled in the second reactor. The effluent from the second reactor was analyzed by three gas chromatographs at regular time intervals. In some experiments, a single-stage fixed-bed continuous flow reactor was also used to evaluate DTP catalysts. In this case, DME, isobutene and nitrogen were used as the reactant.

Results and discussions

Catalytic performance of DTP-1 catalyst

Fig. 2 shows the results of the long run performance test for fresh DTP-1 catalyst and steam dealuminated DTP-1 catalyst. In the figure, the black circle shows the conversion of oxygenates (methanol and DME). Fresh DTP-1 catalyst was deactivated by coke accumulation onto the catalyst surface after 100% of methanol conversion was kept over 700 h of time on stream (**Fig. 2 (a)**). It is well known that the activity of the catalyst deactivated by coke recovers by the burning of the coke. However,

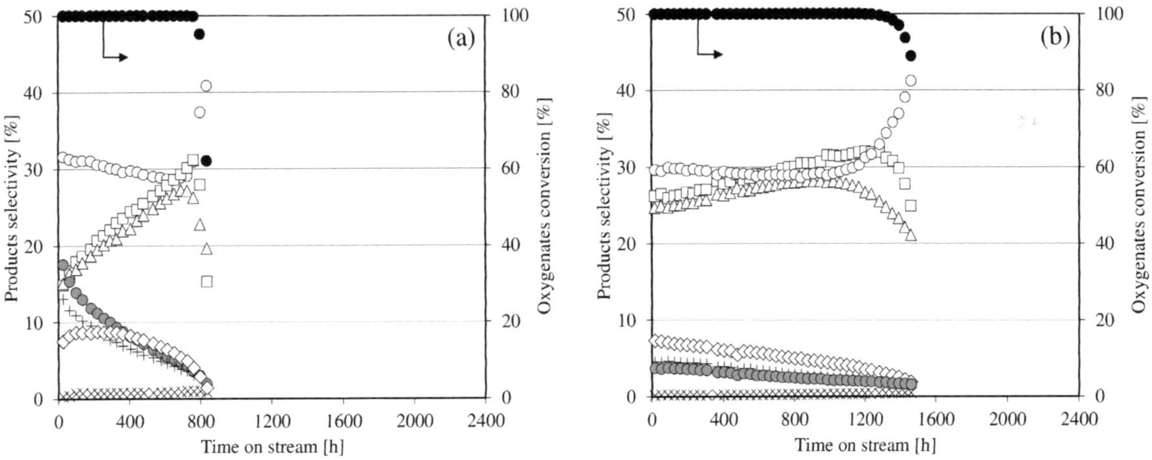

●: Conversion of oxygenates, Selectivity to ◇: ethylene, □: propylene, △: butenes, ×: methane, ●: C_2-C_4 paraffins, ○: C_{5+} and +: BTX.

Catalyst: DTP-1 catalyst, Steam dealumination of catalysts: (a) no steaming, (b) 0.53 MPa at 833K for 50 h.
Reaction conditions: Temperature 753 K, Pressure 0.23 MPa, WHSV of methanol 0.69 h^{-1}, Methanol/isobutene/steam/nitrogen = 1.00:0.24:1.26:0.22(molar ratio).

Fig. 2 Time courses of catalytic performance of fresh and steam dealuminated DTP-1 catalysts.

the catalyst regeneration in commercial units is a cumbersome process, and it would be preferable to decrease the frequency of regeneration.

Additionally, the selectivity to propylene, which is a main product of the process, was low in the initial stage of the run and gradually increased with time on stream, whereas selectivity to aromatics and C_2-C_4 paraffins was decreased with time. **Scheme 1** shows the supposed reaction path for methanol to olefins reaction. C_2-C_4 olefins produced from oxygenates were consecutively converted to undesirable by-products such as aromatics and paraffins with increasing contact time. It is clear from **Fig. 2 (a)** that the contact time of this reaction condition is not the optimum value for the fresh DTP-1 catalyst from the point of view of propylene selectivity.

On the other hand, the selectivity to propylene for the steam dealuminated DTP catalyst in initial stage was higher than that for the fresh DTP-1 catalyst and stable through the run because the progress of consecutive reactions was inhibited by the decrease in the number of active sites on the DTP-1 catalyst (**Fig. 2 (b)**).

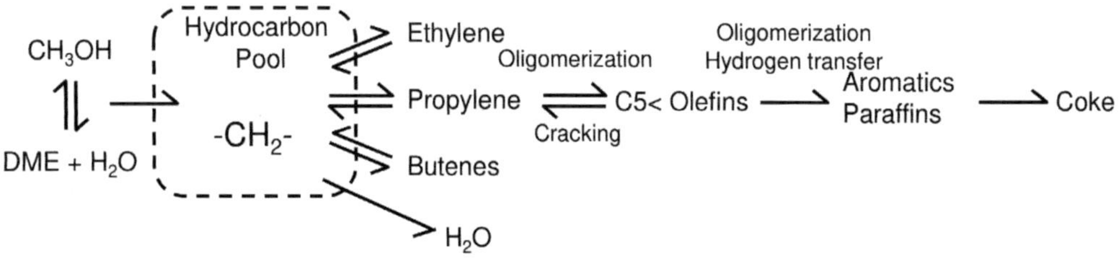

Scheme1. Supposed reaction path for methanol to olefins reaction

Fig. 3 shows the relationships between the contact time and the product selectivity for metal modified ZSM-5 zeolites which have various silica alumina ratios (SAR) in the range of 181-647. The contact time was calculated from not the value of catalyst volume or weight but from the total number of active sites loaded in the reactor. Here we assumed that the number of active sites on each zeolite is equal to the number of aluminium atoms in the zeolite. It was found that the same product selectivity is obtained when the contact time calculated from the number of active sites is the same despite the different acid density for these zeolites. This result clearly shows that the product selectivity strongly correlates with the number of active sites in the reactor in the range of 181-647 of SAR.

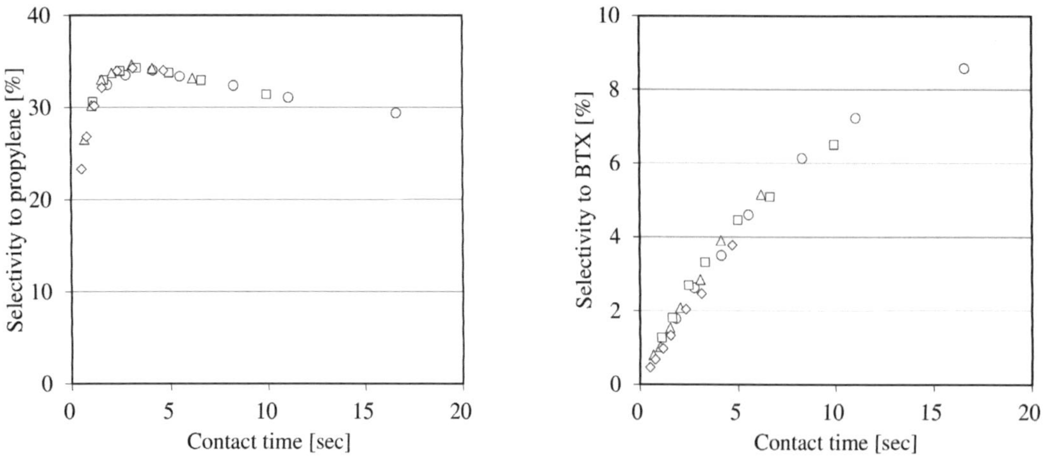

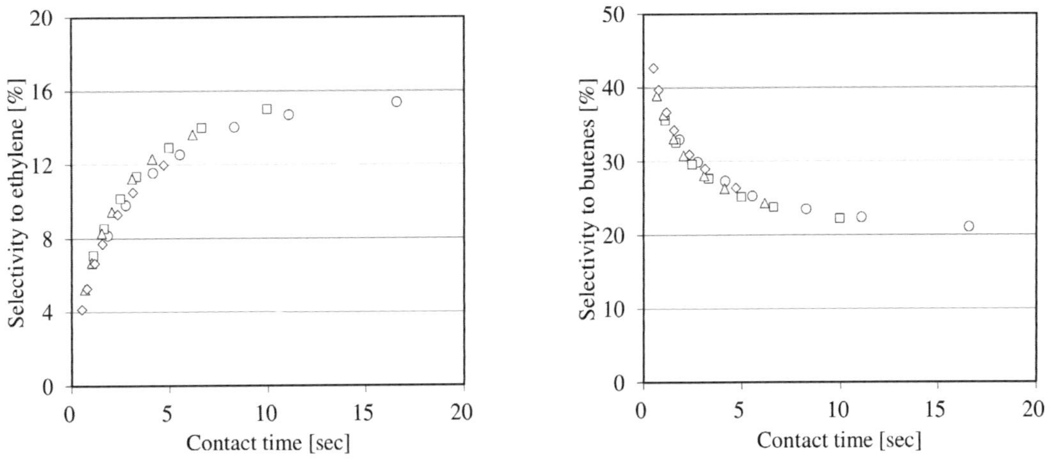

Catalyst: DTP-1 zeolite, Silica alumina ratio: ○181, □348, △490 and ◇647.
Reaction conditions: Temperature 823 K, Pressure Ambient, WHSV of DME 0.36-3.23 h^{-1}, DME/isobutene/ nitrogen = 1.00:4.00:4.00 (molar ratio).

Fig. 3 The relationships between the contact time and the product selectivity for various metal modified ZSM-5 zeolites.

Design concept of newly-developed DTP-2 catalyst

Fig. 4 shows the conceptual scheme of the design of the newly developed DTP-2 catalyst. N_A in the figure represents the initial number of active sites for Catalyst A loaded in the reactor and Life $_A$ represents the life of the catalyst A. N_{MIN} also represents the minimum number of active sites which can keep 100% of oxygenates conversion. Catalyst A is used with the repeated regeneration by coke burning but gradually deactivates by steam dealumination. When the conversion of oxygenates drops below

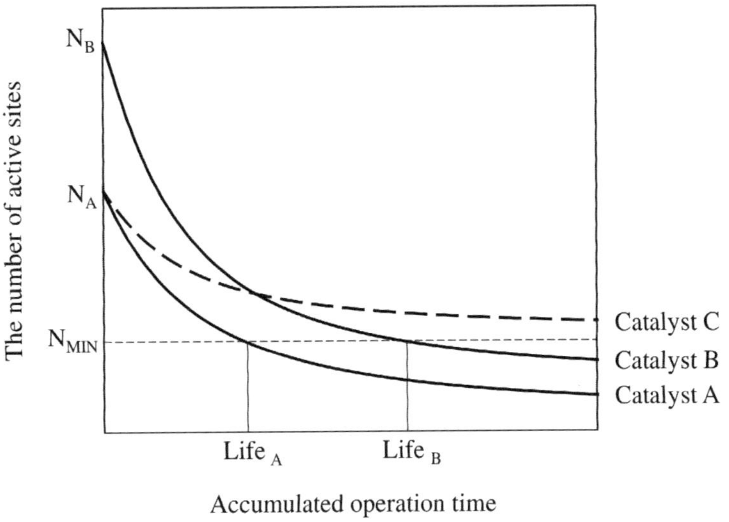

Fig. 4 Conceptual scheme of new catalyst design.

100% due to the decrease in the number of active sites (N_{MIN}), the Catalyst A is removed from the reactor and fresh catalyst will be loaded. Catalyst B is one way to improve catalyst life, that is, if the dealumination rates are same, the catalyst life will be improved with increasing the number of active sites (from N_A to N_B). However, as described above, the product selectivity strongly correlates with the number of active sites and an excessive increase in the number of active sites causes not only the decrease in propylene selectivity but also the need for frequent regeneration as a result of rapid coke formation. **Fig. 5** shows the effect of the SAR of zeolites on the cycle time (the time from the start of reaction to the start of regeneration). With decreases in the SAR of zeolite (with increasing the number of active sites) the cycle time clearly decreases.

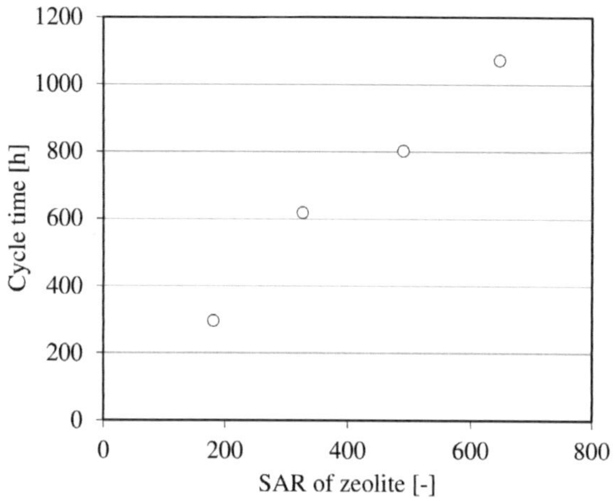

Catalyst: DTP-1 zeolite, Silica alumina ratio: 181-647.
Reaction conditions: Temperature 823 K, Pressure Ambient, WHSV of DME 0.36 h^{-1},
DME/isobutene/ nitrogen = 1.00:4.00:4.00 (molar ratio).

Fig. 5 The effect of SAR to cycle time.

On the other hand, Catalyst C shows our new concept to improve the DTP-1 catalyst; if the rate of dealumination is slowed down, the catalyst life will be improved with high propylene selectivity. Therefore, we have tried to develop the new catalyst which has moderate activity and highly hydrothermal stability.

Hydrothermal stability of DTP-2 catalyst

Both DTP-1 and DTP-2 catalysts were treated under high temperature and pressure steam condition for various treatment times and were subsequently checked for their catalytic performance to investigate the hydrothermal stability of the catalysts. According to **Fig. 3**, product selectivity strongly correlates with the number of active sites on the catalyst. Therefore, the number of remaining active sites after steam dealumination can be estimated from the comparison of the product selectivity of the catalysts. **Table 2** shows the result for the comparison of hydrothermal stability between DTP-1 and DTP-2 catalysts. In **Table 2**, each value was given by indexation using the number of active sites for fresh DTP-1 catalyst as 100. Fresh DTP-1 catalyst had the highest number of active sites, but it fell to 17% of the fresh one after 50 h of steam dealumination. These results are in agreement with the result of

Table 2. The number of active sites on the DTP catalyst after steam dealumination.

Catalyst	Time of steam dealumination			
	Fresh (0 h)	**50 h**	**100 h**	**200 h**
DTP-1	100	17	5	2
DTP-2	29	15	11	6

Indexation using the number of active sites for fresh DTP-1 catalyst as 100.

Fig. 2. DTP-1 catalyst after 200 h of treatment was difficult to keep 100% of oxygenates conversion over a long time. In contrast, fresh DTP-2 catalyst had only 29% of active sites compared to the DTP-1 catalyst, and this catalyst kept 6% of active sites even after 200 h of treatment. These results suggest that the dealumination rate of the DTP-2 catalyst is relatively slower than that of the DTP-1 catalyst.

Advantage of highly hydrothermal stability

Another operational advantage of the DTP-2 catalyst was considered in terms of operation temperature. **Fig. 6 (a)** shows the temperature dependence of equilibrium composition among ethylene, propylene and butenes (1-butene/cis-2-butene/trans-2-butene/isobutene). According to **Scheme 1**, these lower olefins are primary products for the methanol to olefins reaction. If these olefin components formed in the zeolite pore reach their thermodynamic equilibrium composition, the selectivity to propylene increases with the increasing reaction temperature in the temperature range between 700 to 900 K. In practice, as shown in **Fig. 6 (b)**, the selectivity to propylene in the reaction increased with the increasing reaction temperature. This suggests that the DTP-2 catalyst has a potential to obtain higher propylene selectivity than the DTP-1 catalyst since DTP-2 catalyst has highly hydrothermal stability and is applicable to higher reaction temperatures.

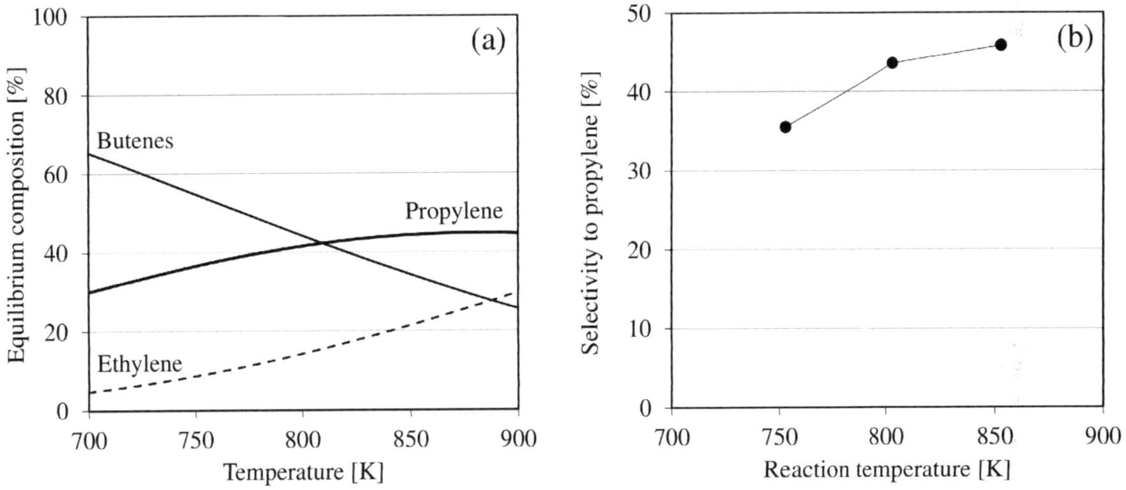

(a): thermodynamic equilibrium composition at 0.1 MPa, (b): result of reaction.
Catalyst: DTP-2 zeolite,
Reaction conditions: Temperature 753-923 K, Pressure A mbient, WHSV of DME 3.0 h^{-1},
DME/nitrogen = 1.00:1.00 (molar ratio).

Fig. 6 Temperature dependence of equilibrium composition among ethylene, propylene and butenes.

Catalytic performance of DTP-2 catalyst

Fig. 7 shows the result of long run experiments for (a) fresh and (b) steam dealuminated DTP-2 catalysts. Fresh DTP-2 catalyst (**Fig. 7 (a)**) kept 100% of oxygenates conversion over 1,000 h; this duration was 1.4 times longer than that for the fresh DTP-1 catalyst. And the fresh DTP-2 catalyst also kept high propylene selectivity through the run although fresh DTP-1 catalyst showed low propylene selectivity in the initial stage of the run. This result is consistent with the result from **Table 2**. Catalytic performance of the steam dealuminated catalyst (**Fig. 7 (b)**) was extremely stable, it kept 100% of conversion over 1,900 h. It was suggested that the DTP-2 catalyst strongly contributes to the stable operation of the DTP reactor for industrial units because product selectivity is very stable through the operation. In addition, both the cumbersome catalyst regeneration process caused by coke accumulation and the exchange frequency of the catalyst caused by irreversible deactivation can be reduced with the use of this catalyst.

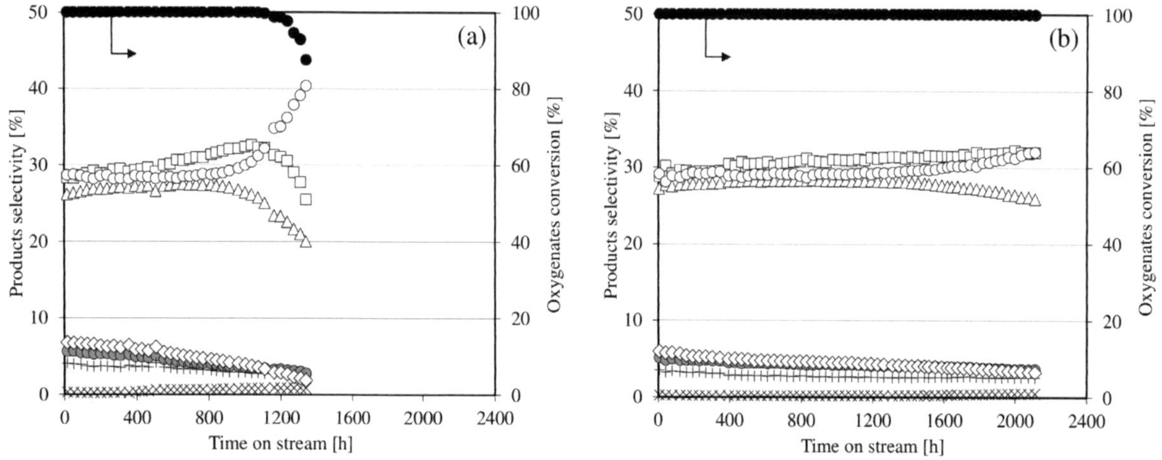

●: Conversion of oxygenates, Selectivity to ◇: ethylene, □: propylene, △: butenes, ×: methane, ●: C_2-C_4 paraffins, ○: C_{5+} and +: BTX.
Catalyst: DTP-2 catalyst, Steam dealumination of catalysts: (a) no steaming, (b) 0.53 MPa at 833K for 50 h.
Reaction conditions: Temperature 753 K, Pressure 0.23 MPa, WHSV of methanol 0.69 h^{-1},
Methanol/isobutene/steam/nitrogen = 1.00:0.24:1.26:0.22(molar ratio).

Fig. 7 Time courses of catalytic performance of fresh and steam dealuminated DTP-2 catalysts.

Summary

A highly propylene selective and stable DTP-2 catalyst which has moderate activity and highly hydrothermal stability has been developed. The new catalyst showed high and stable propylene selectivity through the operation. In addition, it is expected that both the cumbersome catalyst regeneration process and the exchange frequency of the catalyst are reduced through the use of this catalyst.

References

1) Patent WO2005044760.
2) S. Fujita, T. Kanai, Y. Oumi and T. Sano, *Stud. Surf. Sci. Catal.*, **158**(2005)191.

On-Purpose Propylene Production from Light Hydrocarbons with MFI-Zeolite/Metal-Oxide Composites

Dr. Shinya Hodoshima, Azusa Motomiya, Shuhei Wakamatsu and Dr. Fuyuki Yagi*
Research & Development Center, Chiyoda Corporation, Yokohama, Japan
*E-mail: hodoshima.shinya@chiyodacorp.com

Introduction

Propylene, being a raw material in the petrochemical industry, has been increasingly significant because the global demand for propylene has increased at growth rate of 4-5 wt% annually [1]. Since the majority of propylene (ca. 60%) is currently produced as a byproduct by thermal steam-cracking of hydrocarbon feedstock such as naphtha, the thermal crackers can no longer meet the propylene demand. In addition, this technology requires a huge amount of thermal energy (temperature: 850-900°C) as well as its low weight ratio of propylene to ethylene (ca. 0.5). The recent shift to ethane-based feedstock for producing ethylene also might affect the propylene supply, because this technology produces little propylene. Any new method for producing propylene efficiently from light-naphtha, being widely available feedstock, should be thus developed to meet future demand for propylene and reduce energy-consumption in olefin production. Moreover, the conversion of various light hydrocarbons (e.g., shale oil, condensates recovered from shale gas and GTL-naphtha) into valuable light olefins is also significant task in terms of the efficient use of resources.

Catalytic cracking of naphtha-based hydrocarbons over solid-acid catalysts, e.g., ZSM-5, have been actively investigated as an alternative method for on-purpose propylene production to conventional thermal steam-cracking [2,3]. In particular, catalytic cracking using fixed-bed reactor has the following advantages compared with steam-cracking: (1) high weight-ratio of propylene to ethylene; (2) being energy-saving. Furthermore, the cracking process in fixed-bed mode is superior to fluidized catalytic cracking (FCC) processes in terms of operational simplicity in reaction unit. However, catalytic cracking process using fixed-bed reactor has not been commercialized, because practical cracking catalyst with long lifetime, being applicable to fixed-bed reactor, has not been still developed.

From the above-mentioned viewpoints, we have focused on the development of highly active and stable catalyst for light-naphtha cracking in the fixed-bed mode in order to produce propylene efficiently under mild conditions at around 550°C. The proprietary zeolite-based catalysts for cracking, consisting of ZSM-5 zeolites containing Fe and Ga (Fe-Ga-Al-MFI) and metal oxides (e.g., aluminum oxide and silicon oxide) [4], were developed in the present study. This paper presents excellent performance of these proprietary catalysts for cracking.

Experimental procedure

Preparation of zeolite-based catalysts

The desired MFI-type zeolites containing some heteroatoms (Al-MFI(ZSM-5) and Fe-Ga-Al-MFI) were hydrothermally synthesized in autoclave at 120-180°C for 24-72 h under adequate stirring conditions. These sodium-type zeolites were converted into proton-type ones by ion-exchange with NH_4NO_3 aqueous solution and subsequent calcinations at 550°C. The obtained zeolites were characterized by XRD, XRF and SEM measurements. Acid densities of these zeolites (Si/Al and

Si/(Fe+Ga+Al)) were in the range of 19.4-20.0 (ref. **Table 1**). Finally, the resultant powdered-zeolites were pressed, crushed and sorted into sizes of 350-600 μm for reaction tests [4].

The composite catalysts, consisting of the Fe-Ga-Al-MFI zeolites and metal oxides (Al_2O_3 or SiO_2), were prepared at various mixed ratios (Weight ratio of zeolite/metal-oxide: 75/25 and 90/10) as follows. Firstly, the Fe-Ga-Al-MFI zeolites with different Si/(Fe+Ga+Al) ratios (19.4-177.5) were prepared by the above hydrothermal synthesis. Secondly, the powder mixture of these zeolites and metal oxides were formed into cylindrical shape with diameter of 1.0 mm by extrusion. After these samples were converted into proton-type ones with NH_4NO_3 aqueous solution, calcined at 550°C and sorted into length of 1.0-2.0 mm, both H-Fe-Ga-Al-MFI/Al_2O_3 and H-Fe-Ga-Al-MFI/SiO_2 composites were finally obtained as catalysts for light-naphtha cracking.

Catalytic reaction test

Catalytic cracking of *n*-hexane (*n*-C_6H_{14}) as model compound of light-naphtha over MFI-type zeolites (H-Al-MFI and H-Fe-Ga-MFI) only and H-Fe-Ga-Al-MFI/Al_2O_3 composites were carried out for 5 h in a flow-type continuous reactor to measure activities. The *n*-hexane diluted with nitrogen (molar ratio of nitrogen to *n*-hexane: 15/1) were supplied to the reactor at 550°C with the Liquid Hourly Space Velocity (LHSV) of *n*-hexane at 1.0 h^{-1}. After separating products into gaseous and liquid components through vessel equipped with cooling unit, generation rates of gaseous and liquid products were measured to check mass balance. Both catalytic conversions of *n*-hexane and yields of each product were estimated based on both generation rates of products and gas-chromatographic analysis [4].

Catalytic cracking of *n*-hexane without dilution over H-Fe-Ga-Al-MFI/Al_2O_3 composites were performed for 24 h to examine their catalytic performance under practical conditions. The *n*-hexane without dilution was supplied to the reactor at 565°C with the LHSV of *n*-hexane at 5.0 h^{-1}. In addition, long-term cracking reactions with the H-Fe-Ga-Al-MFI/Al_2O_3 composite (Si/(Fe+Ga+Al) ratio: 121.3, (Zeolite/Al_2O_3 weight ratio: 75/25) and the H-Fe-Ga-Al-MFI/SiO_2 composites (Si/(Fe+Ga+Al) ratio: 121.3, Zeolite/SiO_2 weight ratio: 75/25 and 90/10) were also carried out under the same conditions to evaluate their catalyst lifetime. With respect to the H-Fe-Ga-Al-MFI(Si/(Fe+Ga+Al) ratio: 121.3)/SiO_2 composite at the mixed weight ratio of 90/10, a reaction test for cracking was performed for ca. 1,000 h at 565-595°C with the LHSV of *n*-hexane at 4.5-7.0 h^{-1}.

Results and discussion

H-Fe-Ga-Al-MFI zeolite as main component for catalytic cracking

Comparison of catalytic performance between H-Al-MFI(ZSM-5) zeolite and H-Fe-Ga-Al-MFI zeolite with the almost same acid density (Si/T ratio measured by XRF: ca. 20, T=Al, Fe+Ga+Al) was carried out under the same conditions. Both **Fig. 1** and **Table 1** show product selectivity and catalytic activities in cracking of *n*-hexane diluted with nitrogen at 550°C over these zeoiltes synthesized in the present study, respectively. As shown in **Fig. 1**, the product distribution was obviously changed by introducing heteroatom into the H-Al-MFI zeolite as a reference sample. Though the conventional H-Al-MFI exhibited comparatively high selectivity to light olefins, it also gave high selectivity to aromatics. On the other hand, the H-Fe-Ga-Al-MFI remarkably enhanced selectivity to light olefins by suppressing aromatics formation, so that high propylene yields of 27.0 wt% were obtained in spite of low temperature of 550°C for cracking reactions and these activities were kept constant during reaction time (5 h). In general, catalytic cracking of paraffinic hydrocarbons over solid-acid catalysts proceeds via light olefins successively to generate aromatics and selectivity to light olefins strongly depends on acidity of

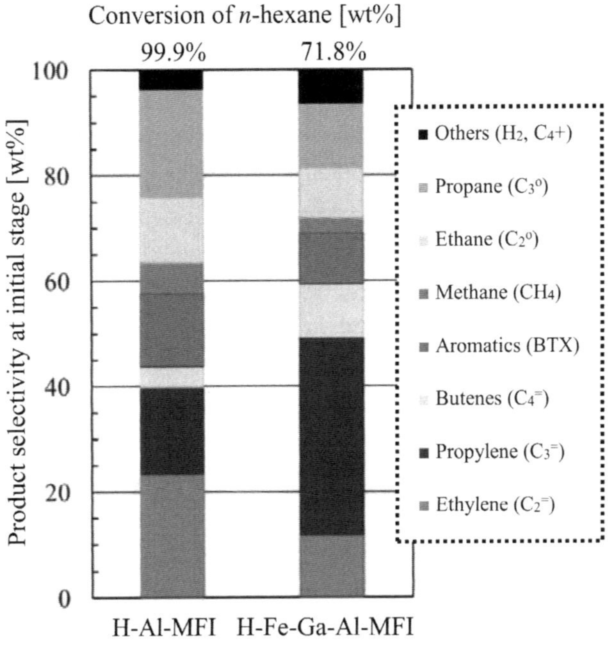

Catalyst(zeolite species);
H-Al-MFI(Si/Al=20.0), H-Fe-Ga-Al-MFI (Si/(Fe+Ga+Al)=19.4, Al/(Fe+Ga+Al)=0.3, Ga/(Fe+Ga+Al)=0.3, Fe/(Fe+Ga+Al)=0.4)
Reaction conditions; Temp.: 550°C, Press.: 0.1 MPa, Cat. vol.: 1.0 mL, Feedstock: n-C_6 diluted with N_2, Molar ratio of n-C_6/N_2: 1/15, LHSV of n-C_6: 1.0 h^{-1}

Fig. 1 Initial selectivity in n-hexane cracking over H-Al-MFI and H-Fe-Ga-Al-MFI zeolites

Table 1. Initial activities for n-C_6 cracking over H-Al-MFI and H-Fe-Ga-Al-MFI zeolites

Zeolite species	Acid density* [mol/mol]	Conversion [wt%]	Yield [wt%] $C_3^=$	Yield [wt%] Aromatics
Al-MFI	20.0	99.9	16.3	14.2
FeGaAl-MFI	19.4	71.8	27.0	7.1

*Si/T ratio of zeolite species (T=Al, Fe+Ga+Al)
Temp.: 550°C, Press.: 0.1 MPa, Feedstock: n-C_6 diluted with N_2, Mixed molar ratio of n-C_6/N_2:15/1, LHSV of n-C_6: 1.0 h^{-1}

Table 2. Catalytic activities for n-C_6 cracking over MFI-zeolite and MFI/A_2O_3 composite

Catalyst	Acid density [mol/mol]	Conv. [wt%]	$C_3^=$ yield [wt%]	Productivity [g-$C_3^=$/g-zeolite h]
FeGaAl-MFI only	19.4	71.8	27.0	0.53
FeGaAl-MFI/Al_2O_3	19.4	80.9	28.3	0.58

Mixed weight-ratio of H-Fe-Ga-Al-MFI/Al_2O_3: 75/25
Temp.: 550°C, Press.: 0.1 MPa, Feedstock: n-C_6 diluted with N_2, Mixed molar ratio of n-C_6/N_2: 15/1, LHSV of n-C_6: 1.0 h^{-1}

catalysts. Since iron atoms and gallium atoms in the zeolite have been recognized to reduce acid strength of zeolites [5] and accelerate dehydrogenation of alkanes [6], respectively, the combination of Fe and Ga atoms was considered to cause a drastic change of acidity. As a result, the H-FeGaAl-MFI-zeolite with optimized composition (Al/(Fe+Ga+Al)=0.3, Ga/(Fe+Ga+Al)=0.3, Fe/(Fe+Ga+Al)=0.4) gave the excellent catalytic performance for light-naphtha cracking compared to the conventional ZSM-5.

H-Fe-Ga-Al-MFI/metal-oxide composites for catalytic cracking

In general, zeolite-based catalysts for industrial use are formed into certain shapes by using binder (e.g., metal oxide) to improve mechanical strength or reduce pressure drop during reaction. In the present study, cylindrical composite catalyst with diameter of 1.0 mm, consisting of H-Fe-Ga-Al-MFI zeolite and aluminum oxide (Al_2O_3) binder, were prepared at the mixed ratio of 75 wt%-zeolite/25 wt%-Al_2O_3 by extrusion and its catalytic performance was examined. **Table 2** compares catalytic activities for cracking of n-hexane diluted with nitrogen at 550°C over the H-FeGaAl-MFI/Al_2O_3 composite (75 wt%/25 wt%) and the H-FeGaAl-MFI only. Both n-hexane conversion and apparent propylene yield were improved by utilizing the composite catalyst. Moreover, weight of propylene that can be generated per one hour with zeolite of one gram, i.e. space time yield of propylene, was also enhanced, suggesting that zeolite species were well dispersed by use of metal oxides as a binder [4].

In terms of highly catalytic performance as well as enhancement of mechanical strength required for industrial use, it was thus confirmed that zeolite species combined with metal oxides were utilized efficiently as catalysts for light-naphtha cracking.

Excellent performance of H-Fe-Ga-Al-MFI/metal-oxide catalysts under practical conditions

For the purpose of applying the present catalysts to practical cracking process, catalytic performance of the composites under severe conditions were examined by performing cracking of *n*-hexane without dilution at 565°C. Catalytic cracking of light-naphtha without dilution is superior to catalytic cracking with diluents such as nitrogen or steam, in terms of both no thermal energy for heating diluents and irreversible deactivation due to dealumination [7]. **Fig. 2** presents the relationship between acid density of H-Fe-Ga-Al-MFI/Al_2O_3 catalyst and activities for cracking of *n*-hexane without dilution. In contrast to the cracking of *n*-hexane diluted with nitrogen (**Tables 1 & 2**), the H-Fe-Ga-Al-MFI/Al_2O_3 catalysts with low Si/(Fe+Ga+Al) ratio (19.4) gave high aromatics yields in the cracking of *n*-hexane without dilution. As amounts of acid sites decreased (i.e., higher the Si/(Fe+Ga+Al) ratios), selectivity to light olefins became higher by suppressing aromatics formation and the highest propylene yield (ca. 18 wt%) was obtained at the Si/(Fe+Ga+Al) ratio of 121.3. In addition, this catalyst gave the space time yield of propylene of ca. 1.1 [g-propylene/g-zeolite h], being ca. twice of the value (0.58) in the case of diluting with nitrogen. At higher Si/(Fe+Ga+Al) ratios than 150 (i.e., smaller amounts of acid sites), yields of light olefins decreased because of lower conversions. In the cracking of *n*-hexane without dilution, aromatization proceeds much easier than cracking of *n*-hexane diluted with nitrogen due to bimolecular reaction of *n*-hexane and/or re-adsorption of light olefins on acid sites, indicating that aromatics would be generated quite selectively in case of using zeolites with high acid densities. In other words, zeolites with low acid densities would give high selectivity to light olefins by suppressing aromatics formation.

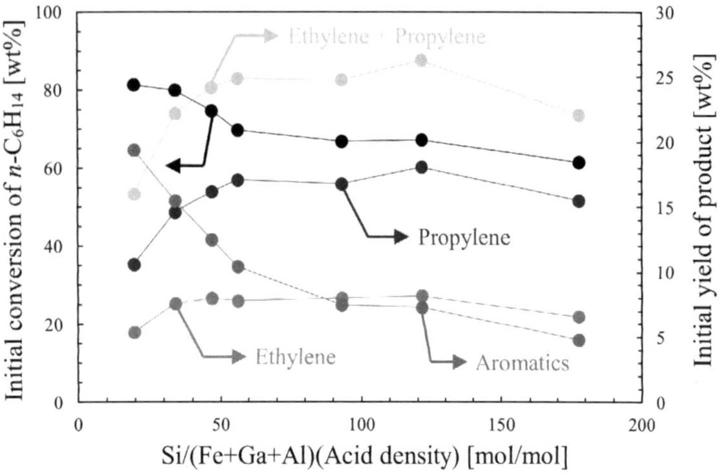

<u>Catalyst;</u>
H-Fe-Ga-Al-MFI/Al_2O_3 (Si/(Fe+Ga+Al)=19.4-177.5 Al/(Fe+Ga+Al)=0.3 Ga/(Fe+Ga+Al)=0.3, Fe/(Fe+Ga+Al)=0.4, Mixed ratio: 75 wt%/25 wt%)

<u>Reaction conditions;</u>
Temp.: 565°C, Press.: 0.1 MPa, Charged volume of catalyst: 2.0 mL, Feedstock: *n*-C_6H_{14} only, LHSV of *n*-C_6H_{14}: 5.0 h^{-1}, Reaction time: 24 h

Fig. 2 Both conversion and product yield in *n*-C_6 cracking without dilution vs. acid density

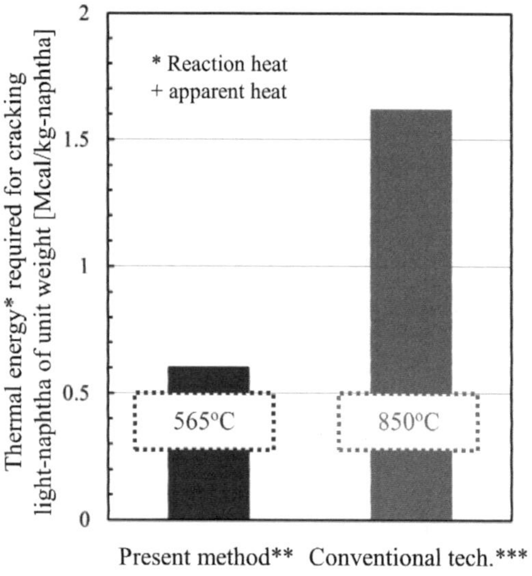

** Catalytic cracking of *n*-C_6 without dilution over H-FeGaAl-MFI/Al_2O_3 (Si/(Fe+Ga+Al)=121.3)
*** Thermal steam-cracking (Non-catalytic)

Fig. 3 Comparison of thermal-energy required for cracking in reaction unit

In order to evaluate the present cracking from the energetic viewpoint, thermal energy required for producing propylene from *n*-hexane was calculated on this experimental basis. **Fig. 3** presents the comparison of thermal energy required for cracking light-naphtha of 1 kg in reaction unit with the conventional thermal-cracking technology. As shown in **Fig. 3**, energy consumption in reaction unit was estimated to be reduced by ca. 60% compared with the thermal-cracking, because no thermal energy for heating diluent such as steam (i.e., apparent heat) was needed as well as its low reaction temperature. The supply of hydrocarbon feedstock without dilution was thus confirmed to be advantageous in terms of energy consumption in reaction unit. In addition, the catalyst lifetime was also investigated by performing *n*-hexane cracking without dilution at 565°C, with respect to the H-Fe-Ga-Al-MFI/Al_2O_3 composite with the Si/(Fe+Ga+Al) ratio of 121.3. **Fig. 4** shows time courses of yields of propylene and ethylene. Propylene yields of ca. 18 wt% obtained at initial stage (ethylene yield: ca. 8.0 wt%) were maintained for ca. 80 h in spite of severe conditions to cause coke formation easily, but it was also suggested that it is necessary to suppress the formation of aromatics to prolong lifetime.

To achieve longer lifetime that is applicable to fixed-bed operation, the H-Fe-Ga-Al-MFI combined with silicon oxide binder, being highly resistant to coke formation, were prepared as cracking catalysts. The H-Fe-Ga-Al-MFI/SiO_2 catalysts were employed for *n*-hexane cracking under the same conditions as the reaction over H-Fe-Ga-Al-MFI/Al_2O_3 composite catalyst. **Table 3** compares catalytic performance for *n*-hexane cracking over the H-Fe-Ga-Al-MFI/Al_2O_3 and H-Fe-Ga-Al-MFI/SiO_2 composites. By use of the H-Fe-Ga-Al-MFI/SiO_2 composite catalyst prepared at the mixed weight-ratio of 75/25, aromatics

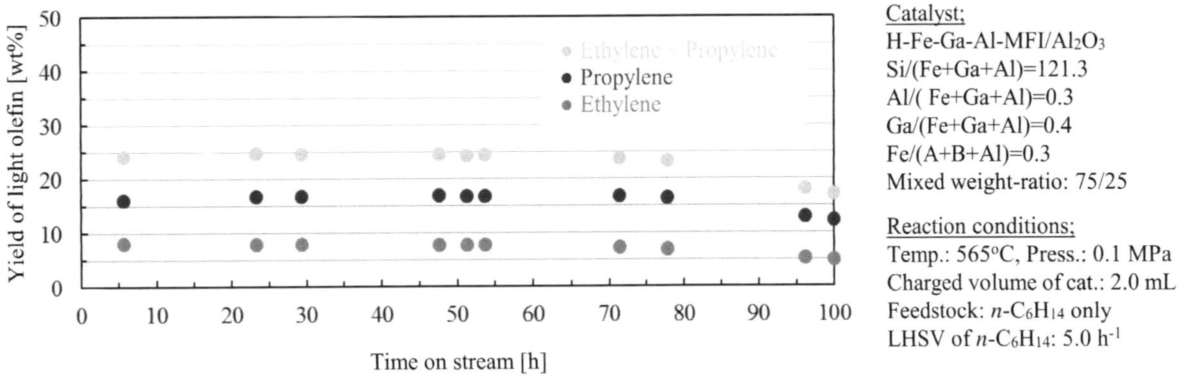

Fig. 4 Time courses of *n*-hexane cracking without dilution over H-Fe-Ga-Al-MFI/Al_2O_3 catalyst

Table 3. Catalytic performance for cracking of *n*-hexane without dilution over H-Fe-Ga-Al-MFI/Al_2O_3 and H-Fe-Ga-Al-MFI/SiO_2 composites

Catalyst species (zeolite(Si/T)/metal-oxide)	Mixed ratio [wt%/wt%]	Conv. [wt%]	Product yield [wt%] $C_2^=$	$C_3^=$	Aromatics	Lifetime of catalyst [h]	Amount of deposited carbon [wt%]
FeGaAl-MFI(121.3)/Al_2O_3	75 / 25	67.3	8.2	18.1	7.3	80	24.6
FeGaAl-MFI(121.3)/SiO_2	75 / 25	64.2	7.4	15.3	4.3	340	4.8
FeGaAl-MFI(121.3)/SiO_2	90 / 10	68.7	8.0	15.6	4.6	480	8.2

Reaction conditions; Temperature: 565°C, Pressure: 0.1 MPa, Charged volume of catalyst: 2.0 mL, Feedstock: *n*-hexane only, LHSV of *n*-hexane: 5.0 h^{-1}

yield was diminished (4.3 wt%) and its activity was kept for 340 h at least, being much longer than that (ca. 80 h) of the H-Fe-Ga-Al-MFI/Al_2O_3 catalyst. Moreover, the catalyst prepared with the mixed weight-ratio of 90/10 also gave lifetime longer than 480 h. Amount of carbon deposits formed on catalyst surfaces in each reaction test are shown in **Table 3**. The H-Fe-Ga-Al-MFI/Al_2O_3 catalyst was deactivated within 100 h (**Fig. 4**) and total amount of carbon deposited on catalyst for 100 h was 24.6 wt%, whereas the H-Fe-Ga-Al-MFI(75 wt%)/SiO_2(25 wt%) catalyst with lifetime longer than 340 h gave total carbon amount of 4.8 wt% for 340 h. The H-Fe-Ga-Al-MFI(90 wt%)/SiO_2(10 wt%) catalyst also exhibited both lifetime longer than 480 h and total carbon amount of 8.2 wt%. It was thus confirmed that catalyst lifetime was closely related to amount of carbon deposit. Since the difference of binder species resulted in remarkable enhancement of catalyst durability, it was considered that a kind of interaction between zeolite and silicon oxide and/or coverage of strong acid sites on external surfaces by silicon oxide gave high resistance to coke formation.

In the cracking of *n*-hexane without dilution at 565-595°C with the LHSV of *n*-hexane at 4.5-7.0 h^{-1} over the H-Fe-Ga-Al-MFI/SiO_2 catalyst, the lifetime longer than 1,000 h was attained, as shown in **Fig. 5**. The achievement of this long lifetime indicates that the catalytic cracking equipped with facilities for regenerating spent catalyst is feasible as an on-purpose propylene-production process in fixed-bed mode (see **Fig.6**). Furthermore, overall propylene yields were estimated to be higher than 20 wt%, being comparable to thermal-cracking or other technologies (e.g., FCC processes).

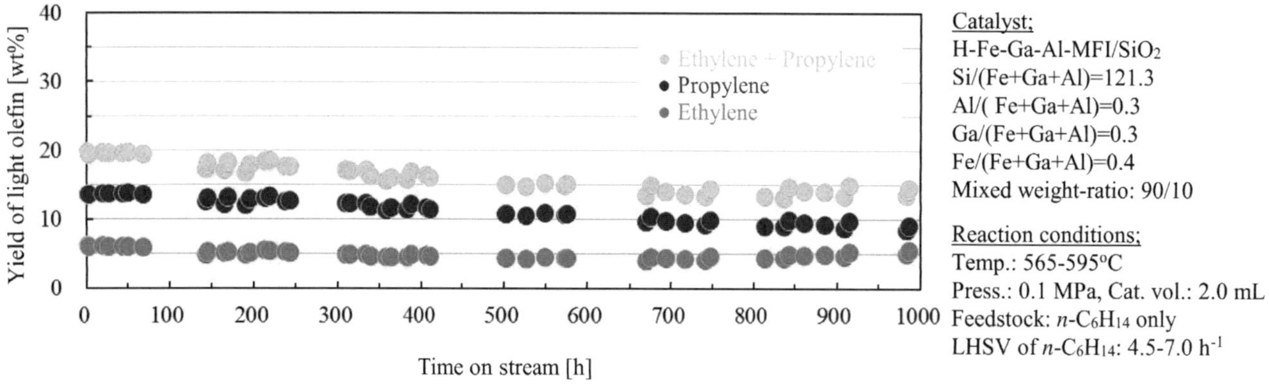

Fig. 5 Time courses of *n*-hexane cracking without dilution over H-Fe-Ga-Al-MFI/SiO_2 catalyst

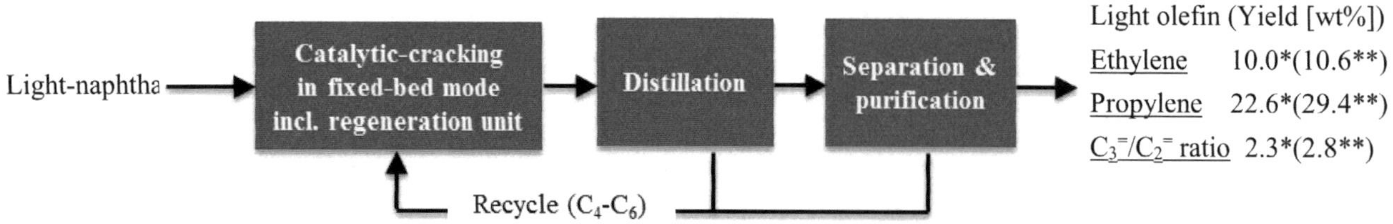

* Catalyst; H-Fe-Ga-Al-MFI(90 wt%)/SiO_2(10 wt%)(Si/T=121.3, Al/T=0.3, Ga/T=0.3, Fe/T=0.4, T=Fe+Ga+Al)
Reaction conditions; Temp.: 565°C, Press.: 0.1 MPa, Feedstock: *n*-C_6 without dilution, LHSV of *n*-C_6: 6.0 h^{-1}

** Catalyst; H-Fe-Al-MFI(68 wt%)/Al_2O_3(32 wt%)(Si/T=31.3, Al/T=0.5, Fe/T=0.5, T=Fe+Al) [8]
Reaction conditions; Temp.: 565°C, Press.: 0.1 MPa, Feedstock: *n*-C_6 diluted with H_2O(25 wt%), LHSV of *n*-C_6: 1.0 h^{-1}

Fig. 6 Simplified process flow for on-purpose propylene production by catalytic cracking of light-naphtha over H-Fe-Ga-Al-MFI/SiO_2 composite

Installation of catalytic cracking in existing naphtha-cracker for increasing propylene yield

As illustrated in **Fig. 6**, the present catalytic-cracking over proprietary zeolites is applicable as an alternative process in simplified fixed-bed mode for on-purpose propylene-production. As well as its excellent stability and low energy-consumption, this cracking process makes it possible to obtain both higher overall yields of propylene than 20 wt% and higher propylene to ethylene ratios than 2.0. It is also noted that various feedstock (e.g., shale oil, condensates recovered from shale gas and GTL-naphtha) are usable for producing propylene as well as light-naphtha.

In addition to the application as a standalone process, an efficient method for increasing propylene yield by integrating this catalytic cracking with existing naphtha-crackers was proposed as another choice for utilizing the present cracking in this paper. **Fig. 7** shows a conceptual process-flow for on-purpose propylene-production by use of binary reaction units for cracking. In this system, reaction unit consists of existing thermal-cracking process and catalytic-cracking unit including distillation equipment for separating unconverted light-naphtha for recycling. Both ethane and propane generated in catalytic cracking are recycled to thermal-cracking reactor through existing equipment for separation and converted into light olefins. Overall ethylene and propylene are obtained as final products through existing facilities for separation and purification. When a certain amount of light-naphtha (100 kg/h) is supplied dividedly to two kinds of cracking reactors at various ratios, overall propylene yield is changed in accordance with the distribution ratio. In case of supplying light-naphtha feedstock of 100 kg/h to catalytic-cracking reactor at X kg/h and thermal-cracking reactor at (100-X) kg/h, respectively, overall yields of ethylene and propylene were calculated regarding the various feed rates (X = 0, 20, 50, 80 and 100 kg/h), on the basis of product yields in each cracking reaction. In this way, the effect on improvement of propylene productivity by installation of catalytic-cracking unit was quantatively evaluated.

Both **Fig. 8** and **Table 4** show both increased ratio of propylene yield and ratio of propylene to ethylene as a function of feed rate of light-naphtha supplied to catalytic reactor. Here the results at the feed rate X = 0 means that this system was operated as a standalone thermal-cracking process, resulting in both ethylene yield of 31.1 wt% and propylene yield of 17.1 wt% ($C_3^=/C_2^=$ ratio: 0.55). By increasing the ratio of light-naphtha supplied to catalytic reactor, overall propylene yield increased, whereas overall ethylene yield decreased, since the present catalytic-cracking gave higher selectivity to propylene ($C_3^=$/

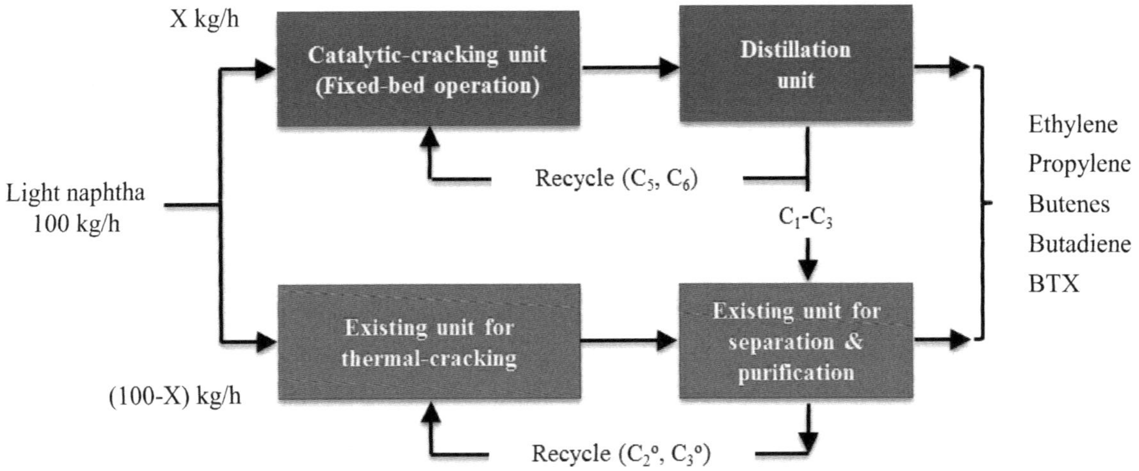

Fig. 7 Conceptual process-design for increasing propylene yield by installing catalytic-cracking unit to existing naphtha-crackers

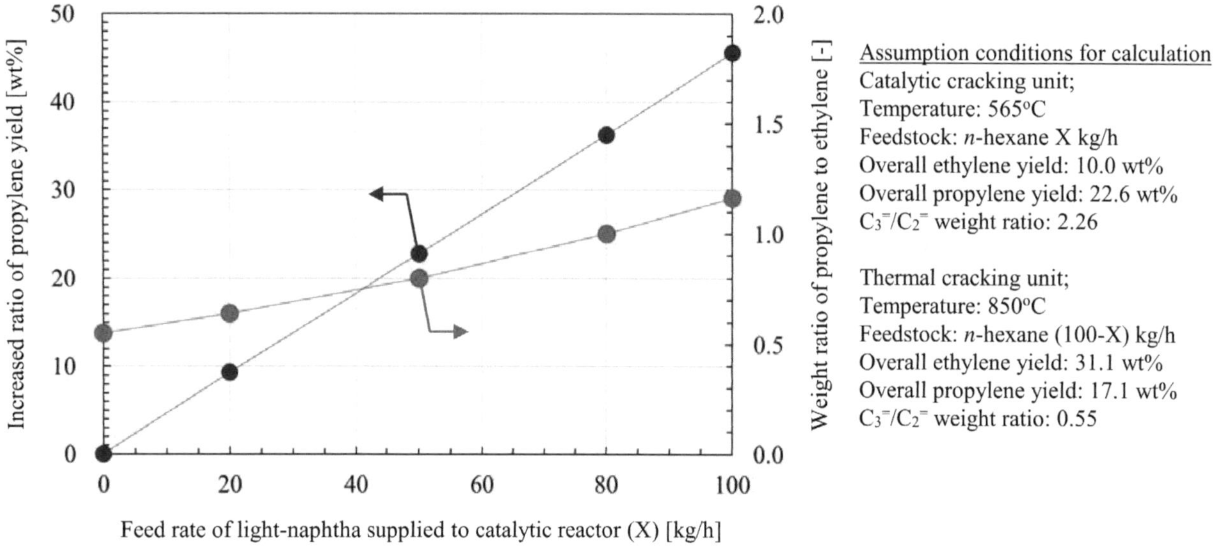

Fig. 8 Improvement of both propylene yield and $C_3^=/C_2^=$ ratio by installing catalytic-cracking

Table 4. Improvement of propylene productivity by installing catalytic cracking unit in thermal steam-cracking technology

Feed ratio of light-naphtha (C.C. [wt%] / T.C. [wt%])	0 / 100 (Thermal cracking only)	50 / 50	100 / 0 (Catalytic cracking only)	
Overall ethylene ($C_2^=$) yield [wt%]	31.1	26.2	21.4	21.9*
Overall propylene ($C_3^=$) yield [wt%]	17.1	21.0	24.9	31.6*
Weight ratio of $C_3^=$ to $C_2^=$ [-]	0.55	0.80	1.16	1.44*
Ratio of increased $C_3^=$ yield [wt%]	-	22.8	45.6	84.8*

Assumption conditions for calculation;
Catalytic cracking unit (565°C):
Feedstock: n-C_6 X kg/h, Overall $C_2^=$ yield: 10.0 wt%, Overall $C_3^=$ yield: 22.6 wt%
* Feedstock: n-C_6 X kg/h diluted with H_2O(25 wt%), Overall $C_2^=$ yield: 10.6 wt%, Overall $C_3^=$ yield: 29.4 wt% [8]
Thermal cracking unit (850°C): Feedstock: n-C_6 (100-X) kg/h, Overall $C_2^=$ yield: 31.1 wt%, Overall $C_3^=$ yield: 17.1 wt%

$C_2^=$ ratio: 2.26) than that in thermal-cracking. In case of supplying all of light-naphtha to the catalytic reactor (X = 100 kg/h), overall yield of propylene was estimated to increase by 45.6 wt% compared with thermal-cracking only (X = 0 kg/h), resulting in both propylene yield of 24.9 wt% and $C_3^=/C_2^=$ ratio of 1.16. Furthermore, by adopting catalytic cracking of light-naphtha diluted steam (25 wt%) [8] under this condition, higher performance (overall propylene yield: 31.6 wt%, $C_3^=/C_2^=$ ratio: 1.44) than catalytic cracking of light-naphtha without dilution was found to be obtained (**Table 4**).

It was thus confirmed from this study that the installation of catalytic-cracking unit to existing thermal-crackers made it possible to increase propylene yield definitely and control the ratio of propylene to ethylene flexibly by changing the ratio of feed-rate to each reactor. In addition, it was also suggested that the load down of the thermal-cracker by introducing the catalytic-cracking unit could reduce total energy-consumption remarkably.

Conclusions

The proprietary composite catalysts, consisting of H-Fe-Ga-Al-MFI-zeolite and metal-oxide such as silicon oxide, developed in the present study made it possible to produce propylene quite efficiently from light hydrocarbons (e.g., light-naphtha) at around 565°C and suppress energy-consumption in reaction unit. It was also confirmed that these catalysts gave long lifetimes that is applicable to cracking process in simplified fixed-bed mode. It was suggested on the basis of these experimental results that both an efficient process for on-purpose propylene-production as an alternative technology to conventional thermal-cracking and an efficient technology for increasing propylene yield definitely by combination with existing cracking technologies are feasible by using the present catalysts.

Acknowledgements

A part of the present study was financially supported by the Japan Oil, Gas and Metals National Corporation (JOGMEC). We would like to express our deep gratitude here. The authors acknowledge Prof. Sachio Asaoka (The University of Kitakyushu) for guidance regarding zeolite synthesis.

References

[1] Forecast of global demand and supply for petrochemical products for the period 2005 to 2018 (http://www.meti.go.jp/press/2014/06/20140613002/20140613002.html)
[2] H. Konno, T. Okumura, Y. Nakasaka, T. Tago, T. Masuda, J. Jpn. Petrol. Inst., **55**(4), 267 (2012).
[3] A. Corma, J. Mengual, P.J. Miguel, Appl. Catal. A; General, **421-422**, 121 (2012).
[4] S. Hodoshima, A. Motomiya, S. Wakamatsu, R. Kanai, F. Yagi, Res. Chem. Intermed., in press.
[5] A. Chatterjee, T. Iwasaki, T. Ebina, A. Miyamoto, Micropor. Mesopor. Mater., **21**, 421 (1998).
[6] T. Inui, F. Okazumi, J. Catal., **90**, 366 (1984).
[7] A. Corma, J. Mengual, P.J. Miguel, Appl. Catal. A; General, **417-418**, 220 (2012).
[8] S. Hodoshima, A. Motomiya, S. Wakamatsu, R. Kanai, F. Yagi, Micro. Meso. Mater., submitted.

401056 Evaluation of Technology Development for Direct Replacement Biofuels Using Thermochemical Processes

Abstract:

Biofuels technology development continues even as ethanol production matured. New "biorefineries" are being commissioned and the U.S. government continues to support development of new technologies and areas to improve overall biofuel production. More conventional biorefineries utilize gasification to convert bioderived feedstocks to fuel. Currently under development are a variety of direct replacement biofuels production technologies that do not gasify biomass.

Evaluating the economics and carbon dioxide lifecycle of "biorefineries" is essential to determine the viability of technologies for the future production of fuels from bioderived feedstocks. Since gasification is an energy-intensive process, it may not be the most economically advantageous biofuel production process. Another process that can convert bioderived feedstocks to a blending fuel is fast pyrolysis, which has many commercial analogs operating throughout the world. With advances in equipment development, fast pyrolysis can be used to directly convert bioderived feedstocks to finished direct replacement biofuels.

Identifying process technology advancements that allow for a wider range of biofeedstocks are considered, in addition to thermochemical conversion technologies and existing patents and work in this area.

Continuous Butanol Production Using an Immobilized-Cell Bioreactor and Online Gas Stripping

Jonathan P. Raftery, Pengji Zhou and Dr. M. Nazmul Karim
Artie McFerrin Department of Chemical Engineering
Texas A&M University, College Station, TX

15 AIChE Spring Meeting, Austin, TX

Artie McFerrin Department of CHEMICAL ENGINEERING TEXAS A&M UNIVERSITY

Energy Production From Biomass

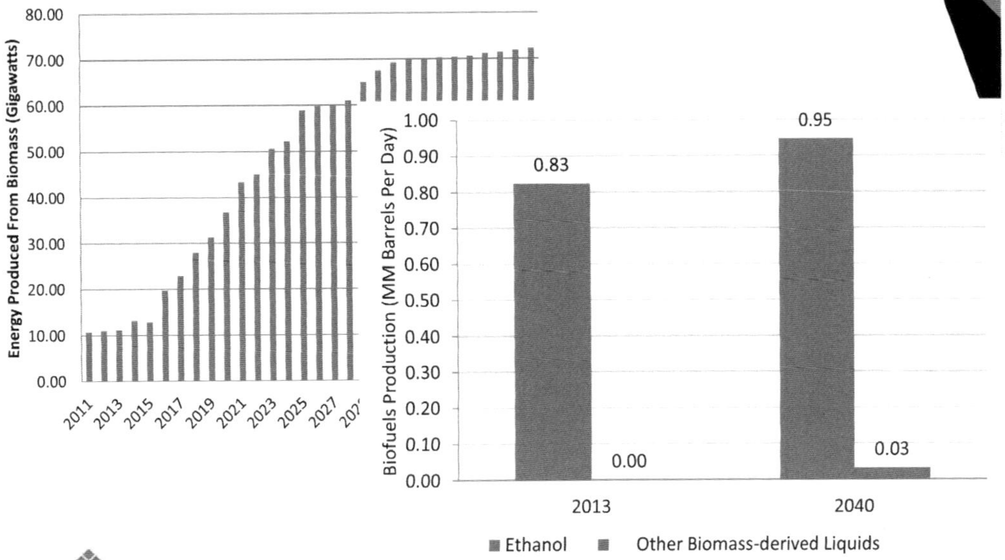

15 AIChE Spring Meeting, Austin, TX

Data from the Energy Information Administration

Artie McFerrin Department of CHEMICAL ENGINEERING TEXAS A&M UNIVERSITY

Butanol Production

- Butanol is a candidate for a promising new biologically-derived fuel additive due to
  - Similar energy density to currently used gasoline blends (29.2 MJ/L vs. 32 MJ/L)
  - Relatively safe to handle due to its lower vapor pressure and volatility as compared to other liquid biofuels such as ethanol
- Acetone, butanol and ethanol (ABE) fermentation is done by the bacteria *Clostridium acetobutylicum*
- Butanol is very toxic to *C. acetobutylicum*, potentially limiting its ability to produce butanol in an economical fashion
- Utilizing a continuous, immobilized-cell ABE reactor with gas stripping can overcome these limitations, but possess additional inherent difficulties

15 AIChE Spring Meeting, Austin, TX

Artie McFerrin Department of CHEMICAL ENGINEERING TEXAS A&M UNIVERSITY

Outline

- Analyze the mass transfer limitations in an immobilized-cell ABE bioreactor
- Develop dynamic models for the steady state, continuous fermentation of ABE
- Construct and optimize a process superstructure including gas stripping for the commercial production of ABE from lignocellulosic biomass

15 AIChE Spring Meeting, Austin, TX

Artie McFerrin Department of CHEMICAL ENGINEERING TEXAS A&M UNIVERSITY

External and Internal Diffusion

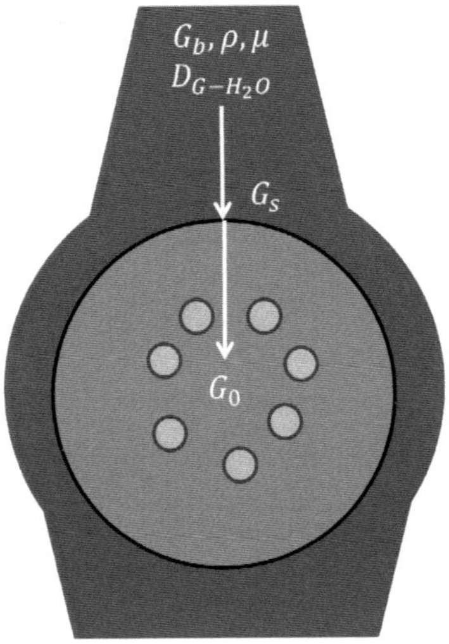

$G_b \rightarrow$ bulk glucose concentration
$G_s \rightarrow$ surface glucose concentration
$G_0 \rightarrow$ center glucose concentration

$D_{G-H_2O} \rightarrow$ glucose/water diffusion coefficient

$\rho \rightarrow$ water density
$\mu \rightarrow$ water viscosity

$d_p \rightarrow$ bead diameter

15 AIChE Spring Meeting, Austin, TX

A&M Artie McFerrin Department of CHEMICAL ENGINEERING TEXAS A&M UNIVERSITY

External and Internal Diffusion

- **External Diffusion**
  - Ranz-Marshall correlation

$$Sh = 2 + 0.6\, Re^{1/2} Sc^{1/3}$$

$$Re = \frac{\rho d_p u}{\mu} \quad \Big| \quad Sc = \frac{\mu}{\rho D_{G-H_2O}} \quad \Big| \quad Sh = \frac{k_c d_p}{D_{G-H_2O}}$$

  - Mass Flux at the Bead Surface

$$W_G = \frac{V_b}{S_b} \cdot \frac{\mu_{max} G_b}{Y_{XG}} = k_c (G_b - G_s)$$

  - Percent of Remaining Glucose (External)

$$\Delta_{ext} = \left(1 - \frac{G_b - G_s}{G_b}\right) \cdot 100\%$$

- **Internal Diffusion**
  - Fick's Second Law for a Sphere

$$\frac{D_{G-H_2O}}{r}\left[r \cdot \frac{\partial G}{\partial r}\right] - \frac{\mu_{max} G}{Y_{XG}} = 0$$

$$@r = 0, \frac{dG}{dr} = 0 \qquad @r = R, G = G_s$$

$$G(r) = G_s - \frac{\mu_{max} G}{6 \cdot Y_{XG}} R^2$$

  - Percent of Remaining Glucose (Internal)

$$\Delta_{int} = \left(1 - \frac{G_s - G_0}{G_s}\right) \cdot 100\%$$

15 AIChE Spring Meeting, Austin, TX

A&M Artie McFerrin Department of CHEMICAL ENGINEERING TEXAS A&M UNIVERSITY

External and Internal Diffusion Limitations

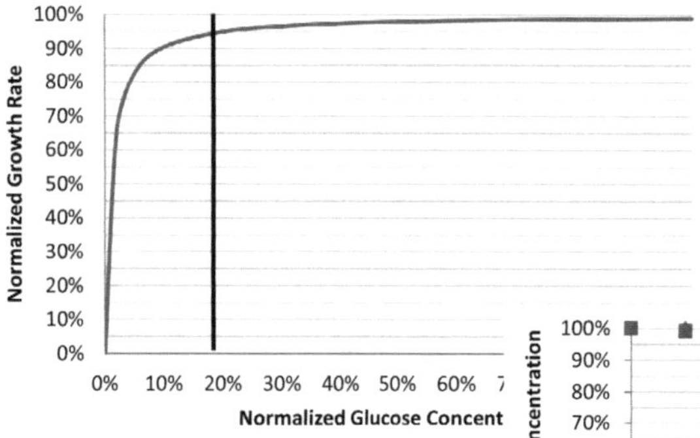

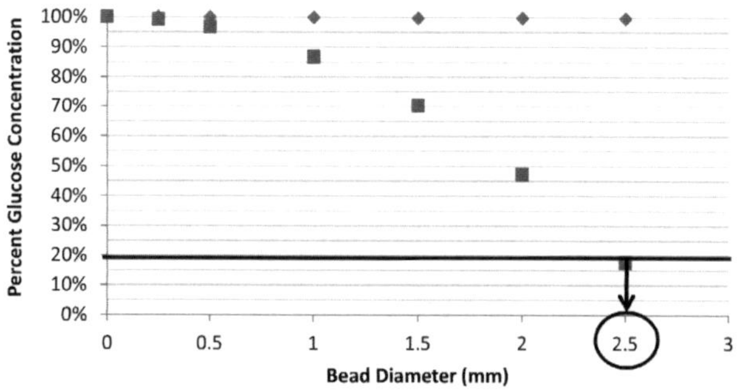

Blue – External
Red – Internal

15 AIChE Spring Meeting, Austin, TX

Artie McFerrin Department of CHEMICAL ENGINEERING TEXAS A&M UNIVERSITY

Continuous ABE Fermentation Modeling

F, G_i

$$\frac{dX_i}{dt} = -k_{loss}DX_i + \mu_i X_i$$

$$\frac{dX_o}{dt} = k_{loss}DX_i - DX_o + \mu_o X_o$$

$$\mu_i = \left(1 - \frac{X_i}{X_\infty}\right) \cdot \left(\frac{\mu_{max}G}{G + K_s\left(1 - \frac{X_i}{X_\infty}\right)}\right) \cdot \left(1 - \frac{B}{B_\infty}\right)^\beta \cdot \left(1 - \frac{A}{A_\infty}\right)^\gamma \cdot \left(1 - \frac{U}{U_\infty}\right)^\delta$$

$$\mu_o = \left(\frac{\mu_{max}G}{G + K_s}\right) \cdot \left(1 - \frac{B}{B_\infty}\right)^\beta \cdot \left(1 - \frac{A}{A_\infty}\right)^\gamma \cdot \left(1 - \frac{U}{U_\infty}\right)^\delta$$

I → acetate
U → Butyrate

$$\frac{dG}{dt} = D(G_i - G) - \frac{1}{Y_X}(\mu_o X_o + \alpha \mu_i X_i)$$

$$\frac{dA}{dt} = -DA + \frac{Y_A}{Y_X}(\mu_o X_o + \alpha \mu_i X_i)$$

$$\frac{dB}{dt} = -DB + \frac{Y_B}{Y_X}(\mu_o X_o + \alpha \mu_i X_i)$$

$$\frac{dE}{dt} = -DE + \frac{Y_E}{Y_X}(\mu_o X_o + \alpha \mu_i X_i)$$

$$\frac{dT}{dt} = -DT + \frac{Y_T}{Y_X}(\mu_o X_o + \alpha \mu_i X_i)$$

$$\frac{dU}{dt} = -DU + \frac{Y_U}{Y_X}(\mu_o X_o + \alpha \mu_i X_i)$$

15 AIChE Spring Meeting, Austin, TX

Artie McFerrin Department of CHEMICAL ENGINEERING TEXAS A&M UNIVERSITY

Continuous ABE Fermentation Results

$\alpha = 0.9$ $\quad k_{loss} = 0.1$

$D = 0.1\ \text{hr}^{-1}$ $\quad G_i = 60\ \text{g/L}$

$@t = 0\ \ X_i = 5\ \text{g/L}$

Component	Steady State Prediction (g/L)
Immobilized Biomass (X_i)	28.946
Free Biomass (X_o)	4.433
Acetone (A)	3.499
Butanol (B)	11.366
Ethanol (E)	0.883
Acetate (T)	2.302
Butyrate (U)	0.965

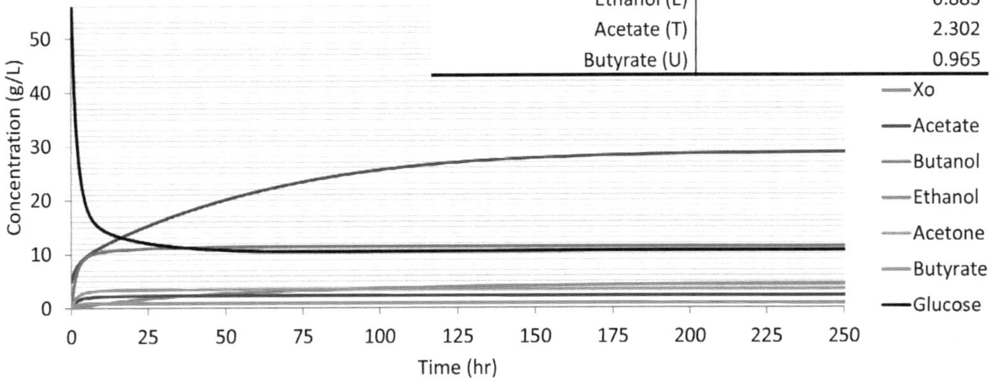

15 AIChE Spring Meeting, Austin, TX

Artie McFerrin Department of CHEMICAL ENGINEERING TEXAS A&M UNIVERSITY

Large-Scale Production of Butanol

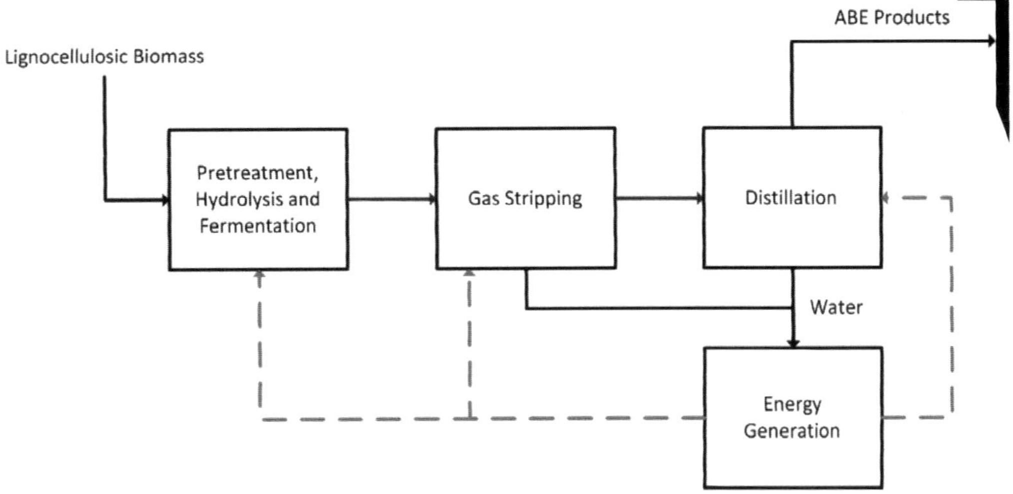

15 AIChE Spring Meeting, Austin, TX

Artie McFerrin Department of CHEMICAL ENGINEERING TEXAS A&M UNIVERSITY

Strategy

- Formulate steady-state models that describe the superstructure
  - Mass balances
  - Energy balances
  - Cost correlations

- Use branch-and-bound optimization techniques to determine the optimal plant topology and operating conditions

15 AIChE Spring Meeting, Austin, TX

Artie McFerrin Department of CHEMICAL ENGINEERING TEXAS A&M UNIVERSITY

MINLP Formulation

$$\begin{aligned} \text{minimize} \quad & f(x,y) \\ \text{subject to} \quad & g(x,y) \leq 0 \\ & x \in X \\ & y \in Y \quad \text{integer} \end{aligned}$$

$f(x,y) \rightarrow$ objective function

$g(x,y) \rightarrow$ process constraints (mass, energy, cost)

$x \rightarrow$ continuous variables (flowrates, etc.)

$y \rightarrow$ integer variables (unit decisions)

15 AIChE Spring Meeting, Austin, TX

Artie McFerrin Department of CHEMICAL ENGINEERING TEXAS A&M UNIVERSITY

Objective Function and Cost Models

$$\min C_t = C_b + C_e + C_a$$

Total Profit Model

$$0 = CC + OC + LC - (P_b \cdot C_b + P_e \cdot C_e + P_a \cdot C_a)$$

Unit Operation Cost Model

$$\text{Cost}_{\text{unit}} = \text{Cost}_{\text{base}} \cdot \left(\frac{\text{Size}_{\text{unit}}}{\text{Size}_{\text{base}}}\right)^{a}$$

$Size_{\text{base}}$ = basis flowrate
$\text{Cost}_{\text{base}}$ = basis cost from literture or Guthrie correlations
$\text{Size}_{\text{unit}}$ = flowrate of designed reactor
$\text{Cost}_{\text{unit}}$ = flowrate of designed unit operation
a = scaling factor, 0.6 used here

15 AIChE Spring Meeting, Austin, TX

Artie McFerrin Department of CHEMICAL ENGINEERING TEXAS A&M UNIVERSITY

Reaction Section

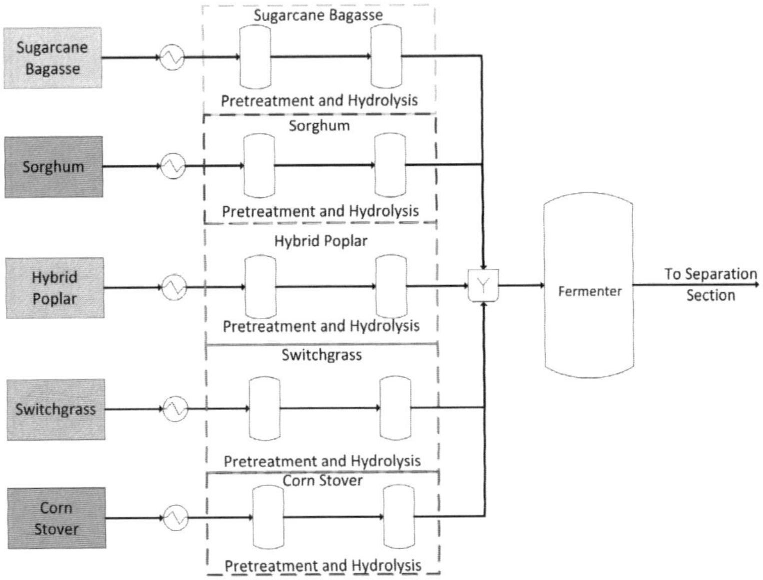

15 AIChE Spring Meeting, Austin, TX

Artie McFerrin Department of CHEMICAL ENGINEERING TEXAS A&M UNIVERSITY

Reaction Section Models

- Pretreatment and hydrolysis based on conversions found in the literature for each biomass/pretreatment combination

- ABE production is based on the stoichiometry of fermentation with *Clostridium acetobutylicum*

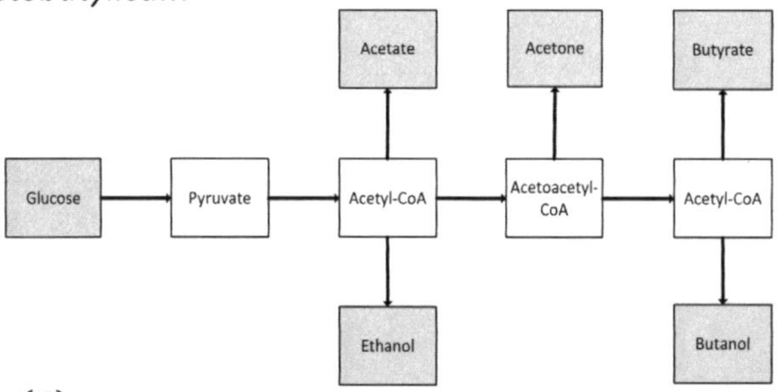

$$100\ C_6H_{12}O_6\ (G) \rightarrow 22\ C_3H_6O\ \ (A) + 56\ C_4H_{10}O\ \ (B) + 7\ C_2H_6O\ (E) +\ 14\ C_2H_4O_2\ (N) + 4\ C_4H_8O_2\ (U) + 211\ CO_2 + 135\ H_2 + C_{41}H_{108}O_{57}\ (X)$$

15 AIChE Spring Meeting, Austin, TX

Artie McFerrin Department of CHEMICAL ENGINEERING TEXAS A&M UNIVERSITY

Gas Stripping Section

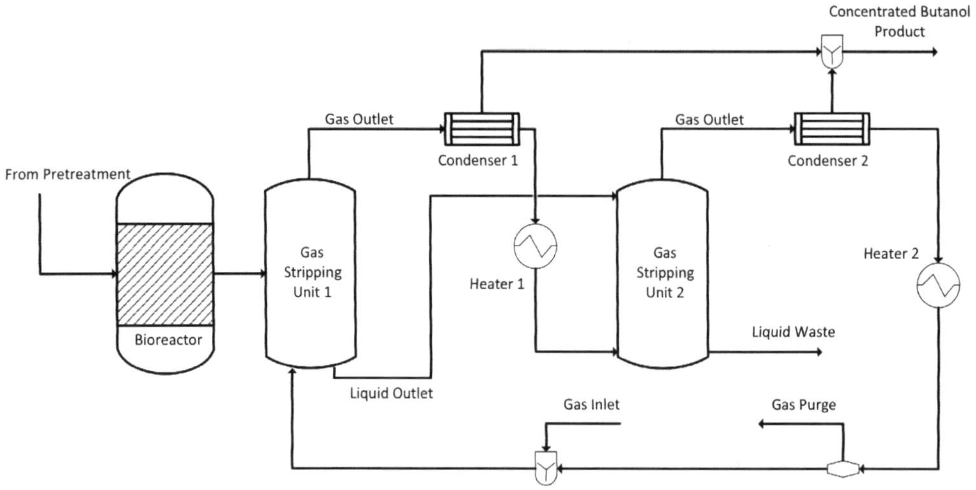

15 AIChE Spring Meeting, Austin, TX

Artie McFerrin Department of CHEMICAL ENGINEERING TEXAS A&M UNIVERSITY

Gas Stripping (Black Box) Models

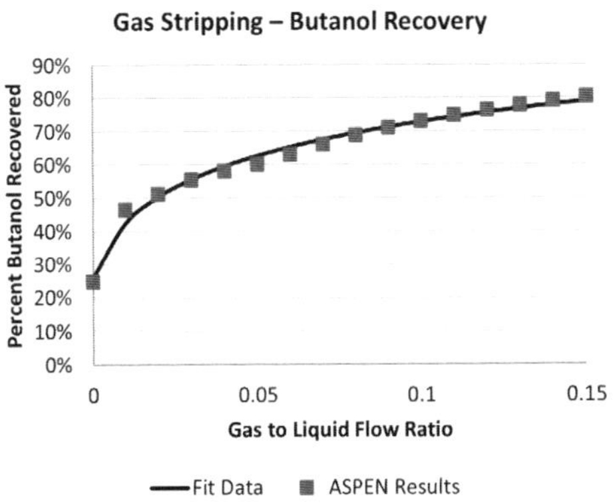

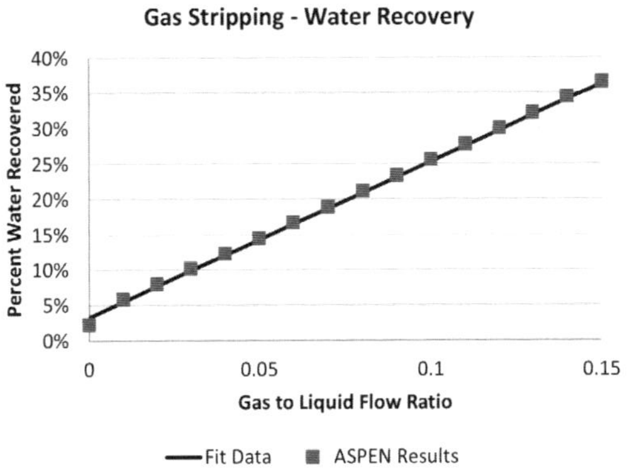

For Ethanol, Butanol, Acetone

$$Recovery_i = A \cdot \log\left(B \cdot \frac{Gas\ Flow}{Liquid\ Flow}\right) + C$$

For Water, Acetate, Butyrate, CO_2

$$Recovery_i = A \cdot \frac{Gas\ Flow}{Liquid\ Flow} + B$$

15 AIChE Spring Meeting, Austin, TX

Artie McFerrin Department of CHEMICAL ENGINEERING TEXAS A&M UNIVERSITY

Gas Stripping Utility Requirement Model

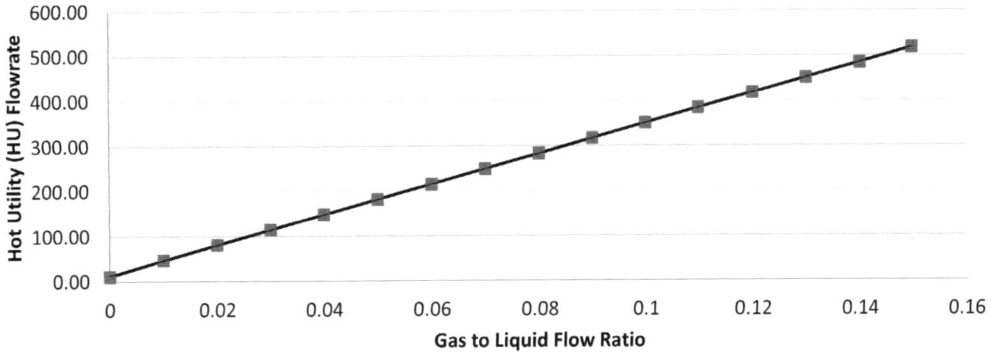

For Heating or Cooling Utility Requirements

$$HU = A \cdot \frac{Gas\ Flow}{Liquid\ Flow} + B$$

15 AIChE Spring Meeting, Austin, TX

Artie McFerrin Department of CHEMICAL ENGINEERING TEXAS A&M UNIVERSITY

Distillation Section

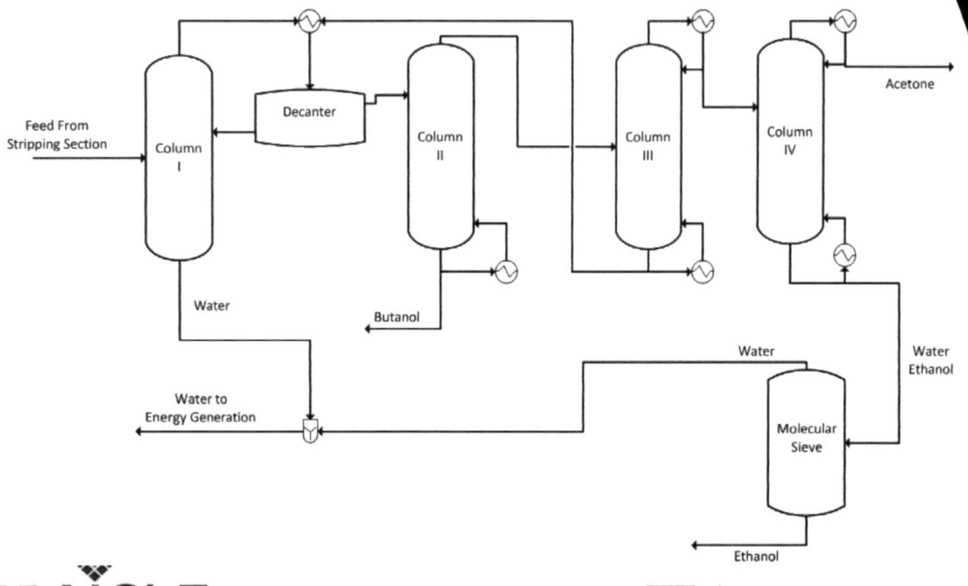

15 AIChE Spring Meeting, Austin, TX

Artie McFerrin Department of CHEMICAL ENGINEERING TEXAS A&M UNIVERSITY

Distillation Models

$$m_j^D = \phi_j \cdot m_j^F$$

$$m_j^B = m_j^F - m_j^D$$

$$Q_{cond} = D \cdot \bar{Q}_{cond}$$

$$Q_{reb} = B \cdot \bar{Q}_{reb}$$

- Models describing a single column
- Use of split fractions (ϕ_j) results in linear models, facilitating the finding of a solution

15 AIChE Spring Meeting, Austin, TX

Artie McFerrin Department of CHEMICAL ENGINEERING TEXAS A&M UNIVERSITY

Other Process Limitations

- Price of acetone

$$P_a \geq \$3.00 \text{ per gallon}$$

- Price of ethanol

$$P_e \geq \$2.25 \text{ per gallon}$$

- Distillation feed composition

$$z_b \geq 0.03$$

15 AIChE Spring Meeting, Austin, TX

Artie McFerrin Department of CHEMICAL ENGINEERING TEXAS A&M UNIVERSITY

Optimization Results – Input & Output

- Biomass and Pretreatment Selection

Sugarcane Bagasse and Sodium Hydroxide

- Biomass Flowrate

3,500 U.S. tons per day (1.2 MM tons per year)

	Optimal Price per Gallon	Market Price per Gallon	Yearly Gallons Produced
Butanol	\$8.98	\$7.94 (2011)	7,972,408.90
Ethanol	\$2.99	\$2.16 - \$2.99	805,986.60
Acetone	\$2.25	\$1.32 - \$2.46	3,533,990.90
Total ABE Production			12,312,386.40

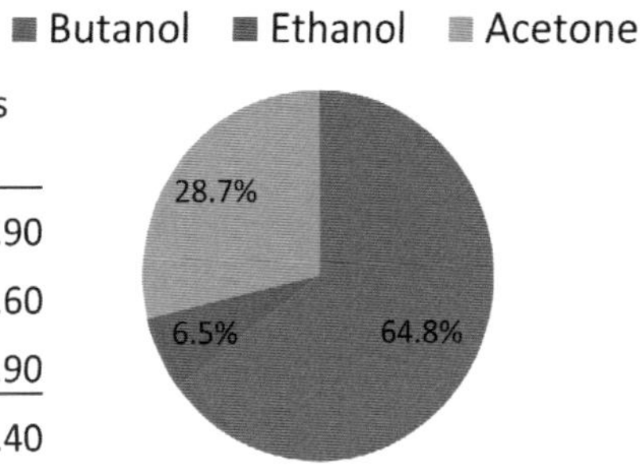

15 AIChE Spring Meeting, Austin, TX

Artie McFerrin Department of CHEMICAL ENGINEERING TEXAS A&M UNIVERSITY

Optimization Results – Gas Stripping

- Stripping gas (CO_2 and H_2) are produced during the fermentation
- No external gas flow rate is needed in the optimal design

	Product Recovery	Reactor Effluent Concentration	Stripping Effluent Concentration
Butanol	21.7%	1.10%	7.10%
Ethanol	26.0%	0.09%	0.70%
Acetone	36.9%	0.40%	3.90%
Acetate	6.7%	0.20%	0.40%
Butyrate	2.4%	0.10%	0.08%
Water	3.3%	88.00%	84.20%

15 AIChE Spring Meeting, Austin, TX

Artie McFerrin Department of CHEMICAL ENGINEERING TEXAS A&M UNIVERSITY

Optimization Results – Process Economics

Assuming a:

3.5% interest rate on capital
Contingency of 20% ISBL
Start-up of 10% total capital
20 year operation lifetime

- Return on Investment
  12.73%
- Discounted Cash Flow Return on Investment
  16.81%
- Payback Period
  ~ 5 years

Capital Cost			$99,284,190.00
	Reactor Cost		$54,125,410.00
	Separation Cost	$	$13,873,290.00
	Heat Exchanger Cost		$31,285,490.00
Operating Cost			$64,964,619.50
	Biomass Cost		$16,669,520.00
	Pretreatment Cost		$1,984,420.20
	Enzyme Cost		$1,121,809.07
	Fermentation Organism Cost	$/yr	$627,191.21
	Process Water Cost		$9,397,119.02
	Stripping Gas Cost		$0.00
	Water Treatment Cost		$29,418,670.00
	Utility Costs		$5,745,890.02
Labor Cost		$/yr	$9,000,000.00

15 AIChE Spring Meeting, Austin, TX

Artie McFerrin Department of CHEMICAL ENGINEERING TEXAS A&M UNIVERSITY

Implications

Sugarcane Production (2011, USDA)

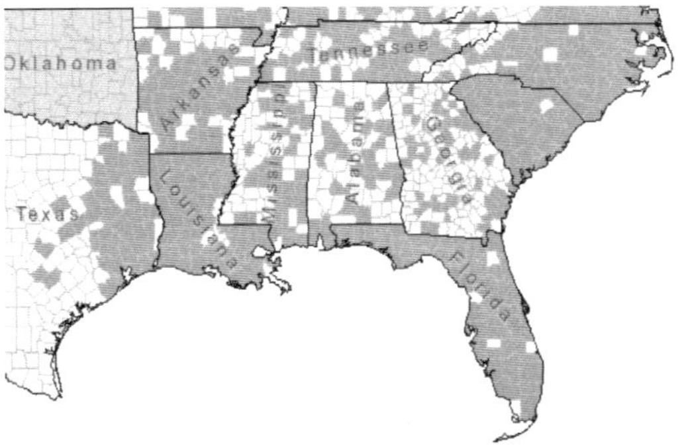

Florida → 13.2 MM tons
Louisiana → 10.8 MM tons
Texas → 1.0 MM tons

- Currently sugarcane bagasse is burned to produce energy for sugar mills
- Repurposing this bagasse in the proposed butanol facility can profitably make a large contribution to the sustainable fuels industry

15 AIChE Spring Meeting, Austin, TX

Artie McFerrin Department of CHEMICAL ENGINEERING TEXAS A&M UNIVERSITY

Conclusions

- Biobutanol is a promising alternative liquid fuel if it can be made at large-scale
- The ABE bioreactor can be designed with negligible mass transfer limitations and can be operated at steady state with a high ABE yield
- Optimization of the steady state operation of a large-scale, continuous biobutanol facility is profitable
- The availability of sugarcane bagasse in Florida, Louisiana and Texas make these preferred locations for biobutanol production

15 AIChE Spring Meeting, Austin, TX

Artie McFerrin Department of CHEMICAL ENGINEERING TEXAS A&M UNIVERSITY

Acknowledgements

Texas A&M Institute for Advanced Studies

National Science Foundation

Dr. Christodoulos Floudas

Karim Research Group

NSF

Texas A&M University Institute for Advanced Study

15 AIChE
Spring Meeting, Austin, TX

Artie McFerrin Department of
CHEMICAL ENGINEERING
TEXAS A&M UNIVERSITY

399679 Lti Model Based Transition Monitoring Applied to Ethylene Plant Dryer Switching

Abstract:

Smooth process transitions during equipment start-ups and shutdowns are very important to maintain process safety and reliability. Equipment changeover transitions are susceptible to faults due to operator variability or equipment failure that, when undetected, can upset the process steady state and decrease the production yield. There are many transition examples in ethylene production where process equipment is taken offline for regeneration or maintenance, and it can be very challenging to isolate transition responses from other miscellaneous disturbances. Operator assisted transitions typically include manual valve operation outside combined with set point changes inside and are accompanied by controller set point and load change responses.

In this work, multivariable transition monitoring is performed by modeling equipment start-up and shutdown transition response steps using linear time invariant (LTI) models. The methodology is tested on a simulated dryer switching procedure in an ethylene plant. Real-time transition monitoring with DCS data could be used to notify operators about transition faults as they occur. We model transition responses with LTI models relating transition variables to valve outputs; however, for manual operations where valve outputs are not available, we use other related readings like pressure and/or flow changes to model the transition responses. Monitoring is performed by calculating the residuals between the LTI model predictions and the actual transition response and using an appropriate fault detection filter.

Big Data Analytics Applied to Real-World Control Systems

From Instruments to Advanced Controls

George Buckbee, P.E.

General Manager, Metso ExperTune

15 AIChE
Spring Meeting, Austin, TX

metso ExperTune

Agenda

- Real-World Data from Control Systems
- Data to Information to Action
- Analysis of a Typical Plant
- Case Study
- Practical Suggestions

15 AIChE
Spring Meeting, Austin, TX

Real-World Data from Control Systems

Good News! You already have the data!

- Thousands of Data Points
  - Controllers (PV, SP, OP, MODE)
- Organized (for Control)
- Somewhat Organized (for Process)
- Historized
  - With Compression?

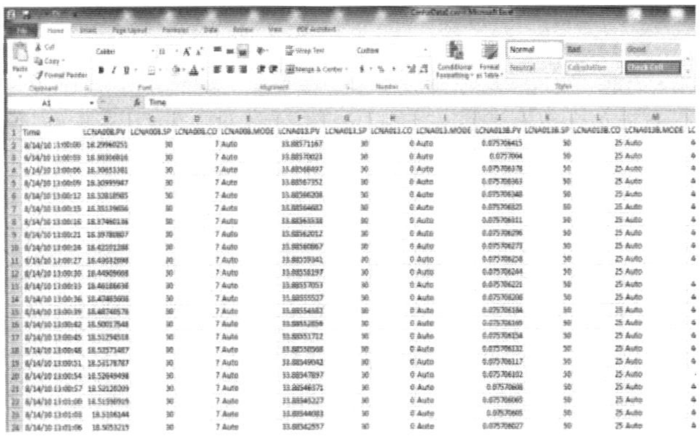

15 AIChE
Spring Meeting, Austin, TX

Data to Information to Action

15 AIChE
Spring Meeting, Austin, TX

Data to Information

DATA

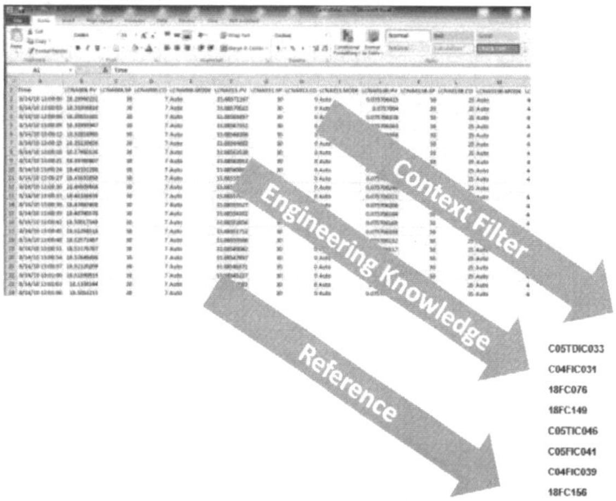

INFORMATION

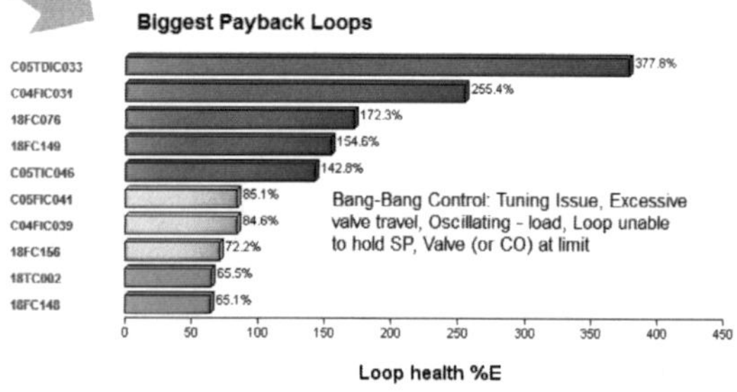

15 AIChE
Spring Meeting, Austin, TX

Simple, Yet Powerful Information from Data

Control Performance Statistics

- % of Time Loop in Manual
- % of Time Valve at Limits
- Statistics
  - Min / Max
  - Avg
  - Std. Deviation
- PV max = PV min

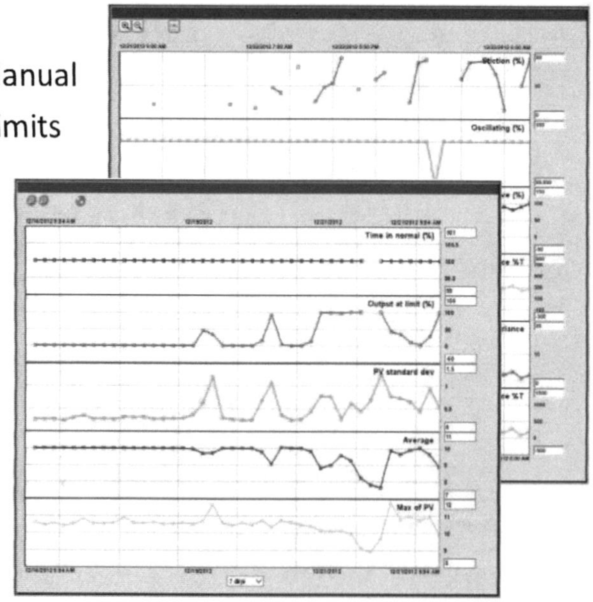

15 AIChE
Spring Meeting, Austin, TX

Automated Data to Information

Control Performance Monitoring

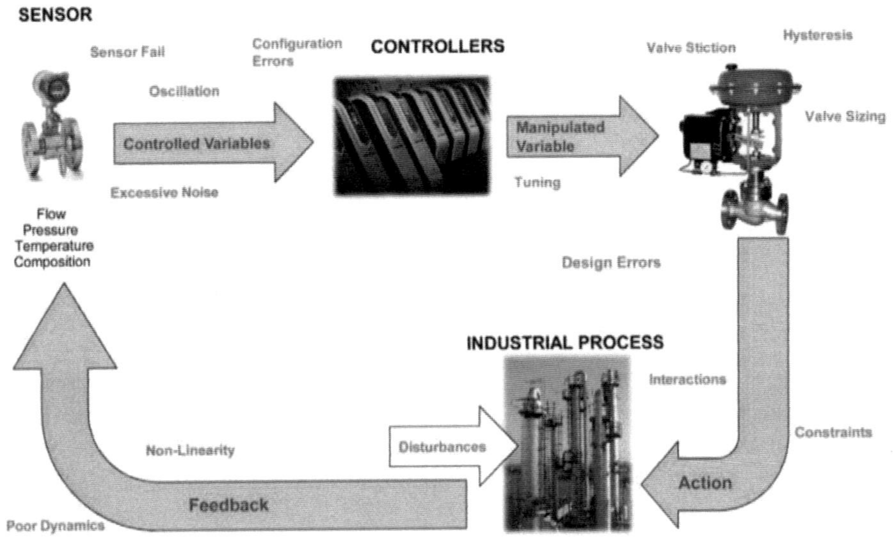

15 AIChE
Spring Meeting, Austin, TX

Example: Data to Information

Measuring Valve Stiction

- Is loop in AUTO?
- Is it Oscillating?
- Does it have a Valve?
- PV = Square Wave
- CO = Saw Tooth
- Peak-to-Trough of CO

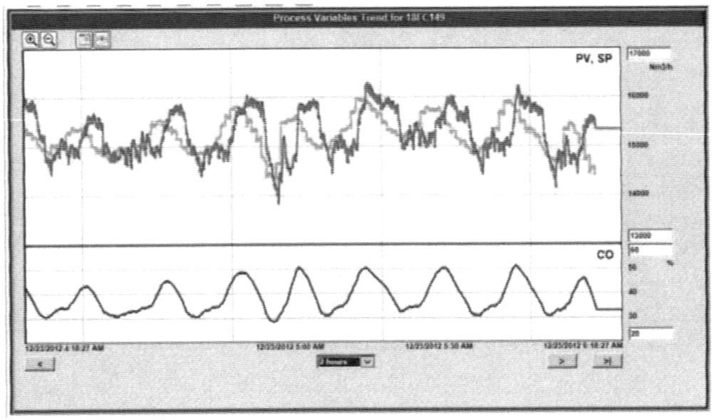

15 AIChE
Spring Meeting, Austin, TX

Existing Control Performance

- Control Performance – The Facts
  - 10%-35% of control loops are in **manual**
  - 1.5M to 5.3M in under utilized assets by running in manual in a 1000 loop site
  - 30% of control valves have problems
  - Are you repairing the correct valves?
  - 30% of control loops are tuned incorrectly, increasing variability in the process.
  - A tremendous amount of money can be saved by understanding the control loop interactions and implementing corrective action.

15 AIChE
Spring Meeting, Austin, TX

Information to Action

INFORMATION

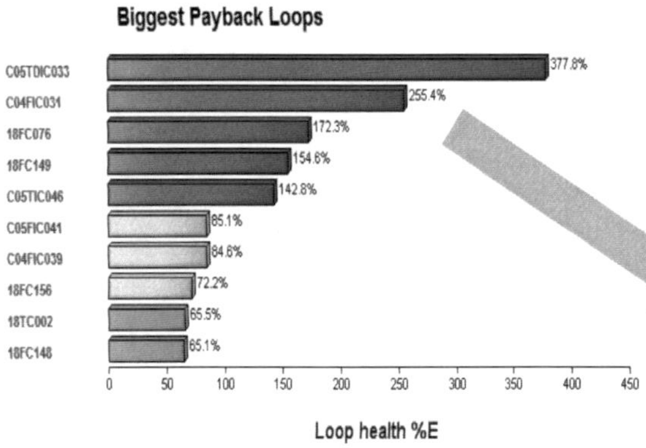

ACTION

15 AIChE
Spring Meeting, Austin, TX

Analysis of a Typical Plant

One-Time Analysis to Identify Opportunity

260 Loops			
% Loops	Issue	Typical	World-Class
46	Loops Unable to Hold SP	1 - 5%	< 1%
89	Loops Not in Normal mode	10% - 25%	< 10%
35	Oscillating	10% - 40%	< 10%
40	Excessive Valve Travel	5% - 15%	< 5%
89	Valve or CO at Limits	5% - 15%	< 5%
53	Valve Stiction	10% - 20%	< 10%
15	Sluggish Tuning	3% - 5%	< 3%
3	Sensor Problems	1%- 3%	< 1%

15 AIChE
Spring Meeting, Austin, TX

Plenty of Data

Distillation column swings

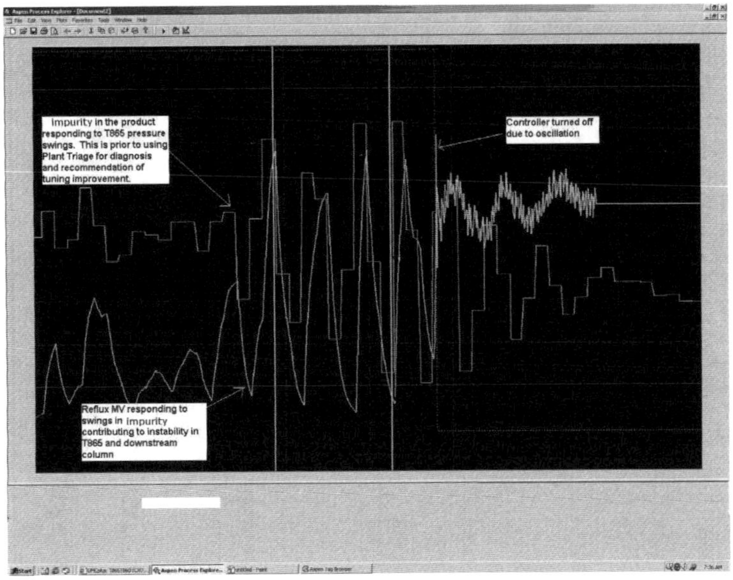

15 AIChE
Spring Meeting, Austin, TX

Use the Data to Obtain Information

Process mass amounts of data into a root cause

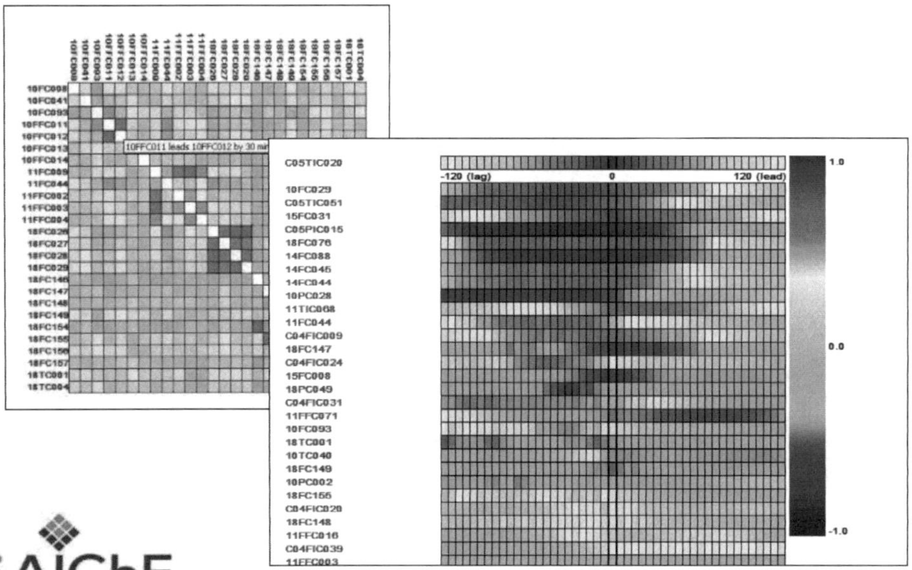

15 AIChE
Spring Meeting, Austin, TX

Take Action

Tune the Pressure controller -use “data” to obtain a process model

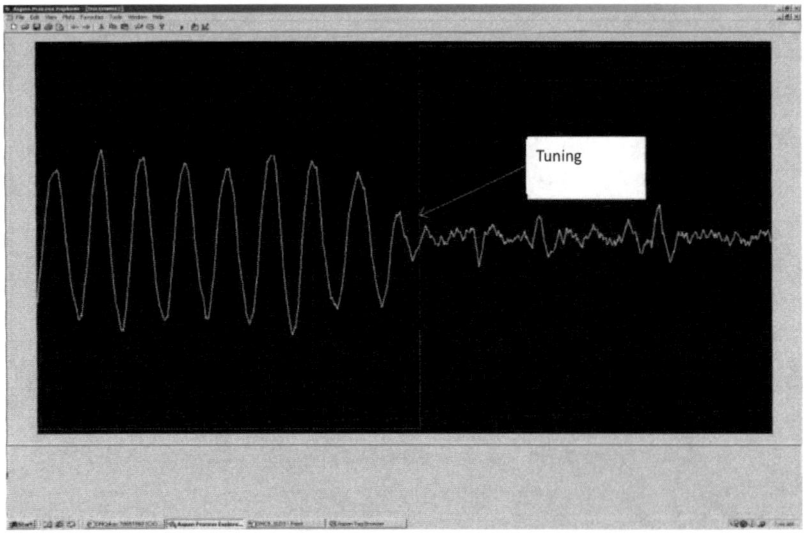

15 AIChE
Spring Meeting, Austin, TX

See Results

More stable and energy savings

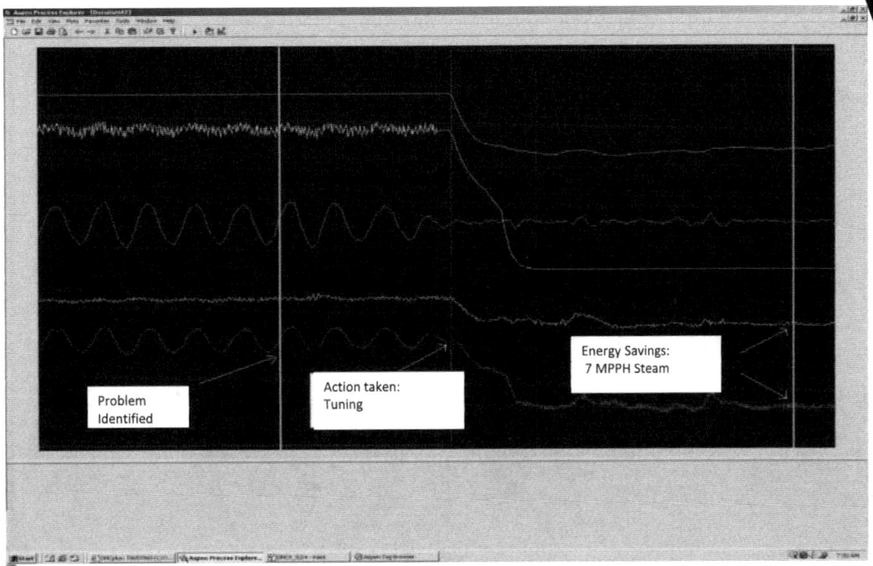

15

15 AIChE
Spring Meeting, Austin, TX

Example: Identify Tuning Opportunities

Active Model Capture

Loop	Description	Harris (normalized) (0-1)	Relative response time (sec)	Model quality	Process dead time (sec)	Process gain	Process lag 1 (sec)	Process lag 2 (sec)
Sort	Sort	Sort ↓	Sort	Sort	Sort	Sort	Sort	Sort
38TC-6046	Темп.разбавительн.	0.9982*	660†	1†	76.52†	0.26†*	84†	0†
2220-LC-3790	L БАК 1ст.СОРТИРОВ./L TANK 1st.SORTIROV .				208.7†	7.2†*	43.2†	0†
2210-TC-3621	Пар из конд.перв.пар/Pairs of kond.perv.par				63.01†	6.8†*	462†	0†
2210-TC-3220	Темп. после нагрева/Temp . after heating	0.9959*	1000†	3†	61.29†	3.4†*	9000†	0†
2220-LV-3955	БАК ФИЛЬТРАТА 233.11/LHC LEACHATE 233.11	0.9947*	2†	1†				
38FC-6127	Поток на входе пресса/Stream entering the press	0.9886*	210†	1†				
38FC-6168	Поток на входе пресса/Stream entering the press	0.9867*	290†	2†				
38FC-6087	Поток разбав.выдувн. бака/Feed razbav.vyduvn . tan	0.9856*	46†	3†				
2210-TCZ-3006	Пар НД в низ бунк./OD pairs in the bottom of his e	0.9856*	620†	2†				

Models for 77 Loops

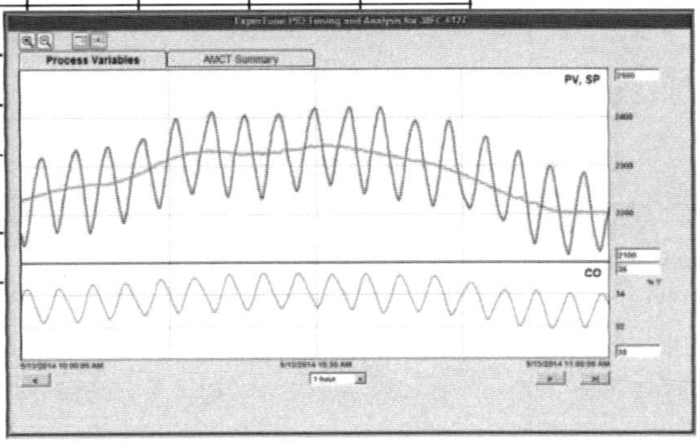

15 AIChE
Spring Meeting, Austin, TX

Practical Suggestions –
An Approach to Control Performance

PLATFORM
- Software
- Hardware
- Installation
- Training
- Site Evaluation

SURGE
- Detailed Unit Evaluation
- Diagnosis
- Corrective Actions
- Value Documentation

SUSTAIN
- Continuous Improvement
- Diagnose & Correct New Issues
- Identify Additional Surges

15 AIChE Spring Meeting, Austin, TX

More Practical Suggestions

- Do a Quick Survey
- Blanket an Area
- Don't be Overly Selective – i.e. Critical Loops Only
- Start with the Basics (Instrument, Valve, Tuning)
- Boost Data Collection Frequencies
- Follow Up is Key
- Learn from Others
- Document the Value

15 AIChE Spring Meeting, Austin, TX

Questions & Answers

15 AIChE
Spring Meeting, Austin, TX

Valve stiction-induced oscillations in chemical process control loops can adversely affect control system performance and degrade the quality of final products. Before stiction can be compensated for, its degree must be accurately estimated. This work proposes a neural network approach to quantify the degree of stiction in a valve once the phenomenon has been detected. Several degrees of stiction are simulated in a closed loop control system by specifying the magnitude of static (fs) and dynamic (fd) friction in a physical valve model. Each simulation generates controller output OP(t) and process variable PV(t) time series data. A feed-forward neural network (the predictor) is trained to model the relationship between a given OP and PV pattern, and the stiction parameters. To test the models predictive capability, a separate set of stiction patterns are generated with added process noise. An inverse neural network-based nonlinear principal component analysis (INLPCA) noise-suppressor effectively extracts the underlying stiction behaviour from the noisy OP and PV stiction patterns. The predictor is shown to estimate fs and fd with a *1.85%* average error.

Control Strategies and Parameters to Improve Reactor Performance

Gregory K McMillan - MYNAH Technologies

Key words: bioreactor control, reactor control, reactor dynamics, reactor optimization

Abstract

There is an incredible difference in process dynamics, control strategies and tuning rules based on the type of reactor. Considerable disagreement has consequently arisen as to the tuning settings and the importance of the dynamics of automation system components. Each control strategy is considered special for the particular production unit and improvements by simple configuration changes are often not recognized.

The type of dynamic response and the source of the dynamics drastically changes with the type of equipment. Reactors can be characterized to having a dead time dominant, moderate (balanced) self-regulating, near-integrating, true integrating, and runaway response. The dead time can range from less than a second to over an hour. These differences have enormous implications as to the importance of automation system dynamics and the best tuning rules and control strategies.

Reactors can be categorized as the type of dynamic response and relative dead time expected based on the reactor type. Dead time dominant responses are rare if you include the dynamics associated the manipulated variable (MV) and the sensor for the process variable (PV) being controlled. Dead time dominance is largely relegated to plug flow reactors with slow at-line analyzers. Normally, plug flow liquid and gas reactors are found to be have a moderate self-regulating response due to heat transfer surface and thermowell lags. Liquid back mixed reactors have a near-integrating, true integrating, and runaway response depending upon degree and sign of internal feedback within the process. Endothermic or moderately exothermic liquid reactors tend to have a near-integrating response (nearly zero process feedback) for continuous operation and a true integrating response (truly zero process feedback) for batch operation. Highly exothermic reactors can have a runaway response (positive process feedback).

The set of tuning rules and performance objectives change as a reactor moves from having a dead time dominant to a moderate (balanced) self-regulating response. The changes correspond to a decrease in negative feedback within the process and possible amplification of noise and interactions.

The objective for dead time dominant and moderate self-regulating process may well be to minimize MV movement and MV overshoot of the final resting value. This is accomplished by tuning rules that minimize proportional and derivative action and maximize integral action.

The objectives for near-integrating, true integrating, and runaway processes are to minimize the peak and integrated error for load disturbances and to minimize overshoot and possibly rise time for setpoint changes. By definition this necessitates significant MV movement and MV overshoot.

Control strategies are presented based on reactor type that will inherently maximize production rate or maintain reaction stoichiometry. The production rate of plug flow gas reactors can be inherently maximized by the temperature controller manipulating reactant feed rates. Reaction stoichiometry for two phase continuous reactors can be inherently maintained by pressure and level control when the product and reactants are in different phases (e.g., gas reactant flow manipulated by pressure control for liquid product flow and gas purge flow). When reactant is recycled, various control strategies are offered to prevent the snowballing effect depending upon the source and path of the recycle.

When maximizing production rate and maintaining reaction stoichiometry is not inherent, control strategies are presented along with PID tuning and the key PID feature required to provide the feedback control needed. The PID feature is found to provide the output response needed to deal with discontinuous signals from analyzers and valves. The feature also enables directional move suppression needed to provide a slow gradual optimization and a fast recovery for fast disturbances and abnormal operation.

Where possible the control strategies are generalized for fed-batch and continuous operations. In cases where the batch response is unidirectional, a PID structure without integral action is used. Alternately, the controlled variable may be translated to the rate of change of a key process variable to enable the use of integral action and the control of the batch profile and end point. A future process variable value is estimated to make decisions as to whether the batch should be terminated or extended.

Reactor Dynamics

There are three fundamental types of process dynamic responses (self-regulating, integrating, and runaway) depending upon the sign of feedback within the process. The sign of the feedback is apparent from whether the open loop response approaches a final value (negative feedback), continues to ramp (no feedback), or accelerates (positive feedback) in the latter portion of the open loop response. The ordinary differential equations for the process as detailed in the 2014 white paper "First Principle Process Relationships" show the first principle origin of the process feedback sign and the terms that define the corresponding open loop response. When the degree of feedback within the process is small within the PID time frame due to a large process time constant, the process can be treated as if there is no feedback.

Insight: Process dynamic responses depend upon the sign and degree of feedback within the process.

Self-Regulating Processes

The most common type of process has internal negative feedback. These processes are self-regulating in that they will settle out at a steady state value as shown in Figure 1 for a given step change in PID controller output assuming there are no process disturbances. Liquid flow, liquid pressure, continuous concentration, and open loop stable continuous temperature processes have a self-regulating response.

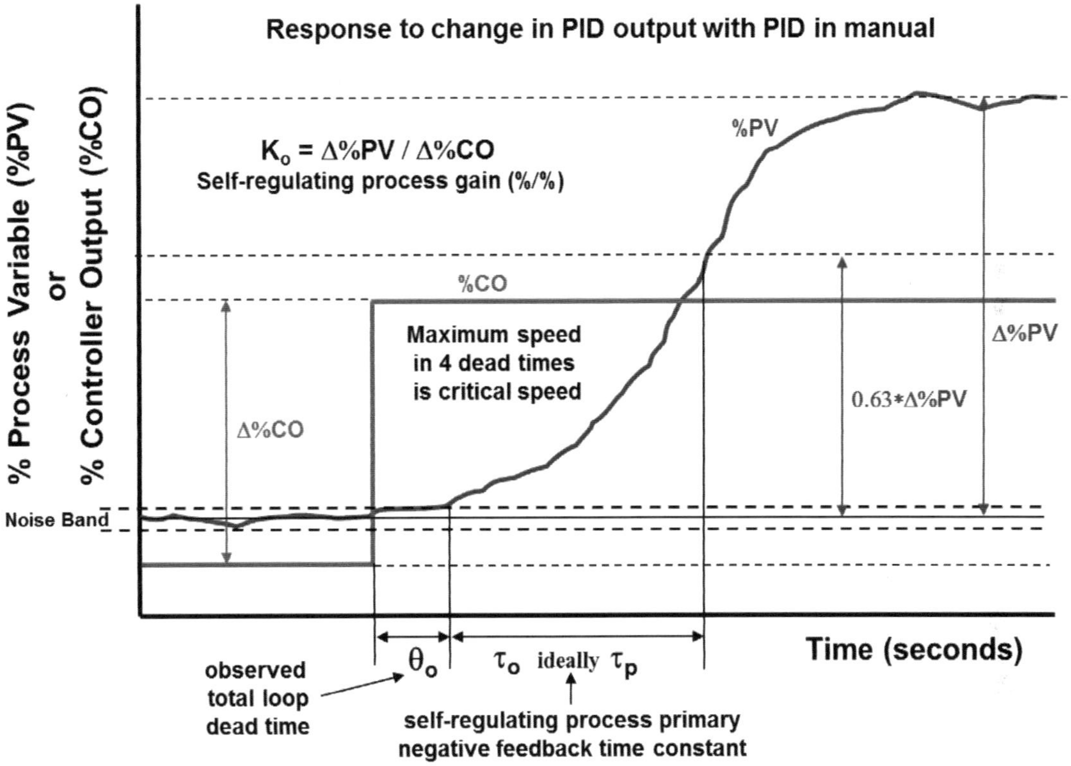

Figure 1 Dynamics for Negative Feedback Process (Self-Regulating Process)

Insight: The most common type of process has internal negative feedback and is called self-regulating because the process will eventually settle out at final value for a given change in PID output.

The primary time constant is the time to reach 63% of the final change in PV value ($\Delta\%PV$) after the PV starts to change for a given change in the controller output ($\Delta\%CO$). The primary time constant is really the largest time constant in the loop. Ideally, the largest time constant is in the process but for liquid flow, pressure, and some inline processes, the largest time constant originates in the sensor, transmitter, or PID filter. Some software can identify the second largest time constant, which is called the secondary time constant. This identification can help in the tuning of controllers that use derivative action because setting the rate time equal to the secondary time constant cancels out its detrimental effect in slowing down closed loop control (control when the

PID is in AUTO, CAS, or RCAS modes). If the process response had no secondary time constant, the slope of the response would be at its maximum value immediately after the dead time. The secondary time constant creates a gradual increase (e.g., bend) in the rate of change of the process after the dead time. All processes have a secondary time constant. If this time constant is not identified, the secondary time constant is converted to an equivalent dead time to provide a first order plus dead time (FOPDT) model. If the secondary time constant is identified we have a second order plus dead time model.

If the primary time constant is less than ¼ the total loop dead time, we have a sub category of a self-regulating process called a dead time dominant process. The process response to a step change in a PID output is essentially a step change in the process variable. These processes are more vulnerable to process noise due to lack of the filtering action within the process. The tuning of such processes requires significantly more integral and less proportional action than all other processes. Derivative action is not used unless there are multiple time constants and no noise.

Insight: A self-regulating process with the primary process time constant less than ¼ the dead time is termed dead time dominant. These processes benefit from less proportional action and usually no derivative action but more integral action.

If the process time constant is greater than 4 times the total loop dead time, we have a sub category of self-regulating process called near-integrating because of the degree of negative feedback within the time frame of the PID. The tuning for near-integrating processes is the same as for true integrating processes, which have zero internal process feedback.

The near-integrating process classification is based on the realization that a PID does most of its correction in the first four dead times during which the process has not appreciably decelerated. Concentration, pH, and temperature control of continuous unit operations with back mixed liquid volumes from boiling, reflux, recirculation, and agitation have a near-integrating response. Some examples are distillation columns, crystallizers, continuous stirred tank reactors (CSTR), evaporators and neutralizers.

Insight: A self-regulating process with a primary process time constant greater than 4 times the dead time is termed near-integrating. Many of the important control loops on liquid unit operations have a near-integrating response. These processes benefit from more proportional and derivative action and less integral action.

When the time constant is between ¼ and 4 times the dead time, we have a moderate self-regulating process. These primarily occur in the gas unit operations in the front end of refineries and hydrocarbon processes. Most of the control literature on the design and implementation of PID, Model Predictive Control (MPC), or special control algorithms addresses moderate self-regulating processes.

Insight: Most of the control literature on PID control is based on a moderate self-regulating process where the primary process time constant is between ¼ and 4 times the dead time.

The open loop steady state gain is dimensionless and is the final change in percent PV for the step change percent change in PID output. Note that the changes in PV and must be in percent of input and output scales because the PID algorithm works in percent and not engineering units.

Integrating Processes

The next most common type of process has zero internal process feedback. These processes are integrating and will ramp continually off to an equipment or process limit as shown in Figure 2 for a given step change in the PID output. Level, gas pressure, and batch concentration, pH, and temperature processes have an integrating response.

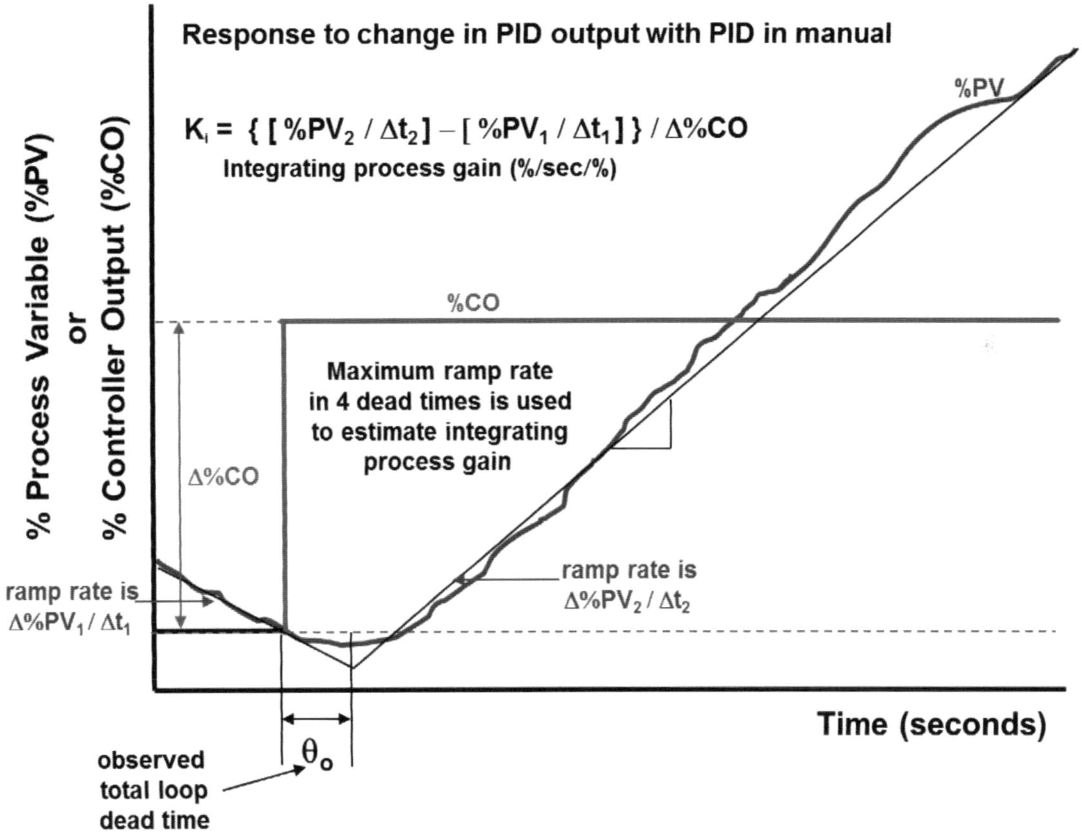

Figure 2 Dynamics for Zero Feedback Process (Integrating Process)

The key characteristic of a zero feedback process is an open loop integrating process gain that is the change in ramp rate of the PV in percent per second ($\Delta\%PV_2 / \Delta t - \Delta\%PV_1 / \Delta t$) divided by the step change in percent controller output ($\Delta\%CO$), assuming there are no process disturbances. If the PID was in AUTO, CAS, or

RCAS modes immediately before the test, the initial ramp rate is typically zero ($\Delta\%PV_1 / \Delta t = 0$). The ramp rate of most integrating processes is extremely slow (e.g., 0.000001 %/sec).

The open loop integrating process gain can be converted to a primary time constant and open loop steady state gain of a near-integrating process and vice versa via Equation 1.3a. The conversion is useful for gaining a better understanding and a set of tuning settings. In fact there are few really true integrating processes. If there were no equipment, process, or measurement limits nearly all integrating processes would reach a steady state. For example if the liquid level or gas pressure in a vessel could increase without bounds, the additional head or pressure would increase the liquid flow or gas flow out of the vessel to balance the increase in flow into the vessel.

Insight: Level, gas pressure, and batch composition, pH and temperature are integrating processes.

An appreciable time constant in a true integrating process is called a secondary time constant. The secondary time constant causes the process response to slowly accelerate to reach the maximum rate of change determined by the open loop integrating process gain. The identification of the secondary time constant and mitigation of the consequences from the lack of inherent deceleration by the use of a rate time equal to the secondary time constant is more important in integrating processes than in moderate self-regulating processes. This definition of secondary time constant is consistent for all three major types of processes enabling the term primary time constant to be used to provide the conversion between a true integrating and a near-integrating response.

To prevent slow rolling oscillations, the product of the PID gain and reset time must always be greater than a minimum established by the open loop integrating process gain defined by Equation 1.3e. Most integrating processes have a gain setting much lower than is permitted by the process dynamics. Since people associate oscillations with too high of a gain and decrease the gain, they make the problem worse when the solution was to increase either the gain or the reset time. Most PID controllers on integrating processes that do not use tuning settings identified by software have too much integral action. For large liquid volumes the software identified setting for PID gain was uncomfortably large. In these cases the gain setting was consequently significantly reduced thinking this makes the loop less likely to oscillate when in fact the loss of negative feedback control action is detrimental to a process without internal negative feedback. The solution would have been to increase the reset time by the same factor to keep the product of the PID gain and reset time the same as computed by the software. The first thing to try in most cases if the period of the oscillation is more than 10 times the dead time is to simply increase the reset time by a factor of 100 to 1000. Whereas decreasing or increasing the PID gain can cause slow or fast oscillations, respectively, increasing the reset time will always make the process more stable.

Insight: Processes without appreciable internal negative feedback have a window of allowable PID gains where a gain too small and a gain too large will cause slow and fast oscillations, respectively.

Insight: PID controllers on integrating processes that are tuned manually generally have a reset time that is orders of magnitude too large, causing slow rolling oscillations.

Runaway Processes

Processes with positive feedback fortunately do not occur very often but are associated with some extremely important unit operations. These processes have an open loop runaway response where the PV will continue to accelerate as shown in Figure 3 for a given step change in the PID output. These processes are called open loop unstable.

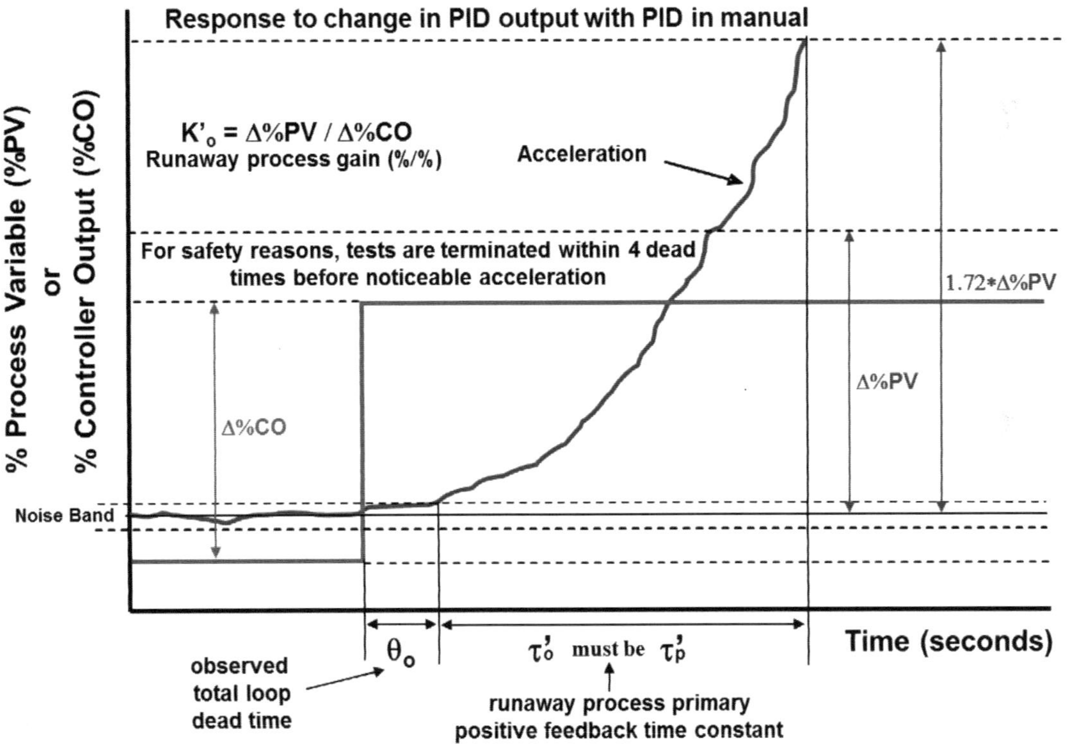

Figure 3 Dynamics for Positive Feedback Process (Runaway Process)

The most common example is an exothermic reactor where the heat of reaction increases with temperature and can exceed the cooling rate if the temperature gets high enough. A much less common example is a runaway speed condition when a large compressor goes into surge when an axial compressor has extremely low rotor inertia. The cell mass concentration control of a biological reactor can exhibit a positive feedback response in the exponential growth phase where an increase in cell concentration increases the cell growth rate. A runaway condition in such a reactor is not typically a concern because the acceleration in the cell mass concentration is so slow and is followed by a stationary phase where the acceleration stops due to the death rate of older cells equaling the growth rate. The pH response of a true strong acid and strong base neutralizer will dramatically accelerate as the pH approaches neutrality (the rate of change of pH increases by a factor of ten for each pH unit in the approach to neutrality). Fortunately most systems are moderated by slight concentrations of weak acids and weak bases, most notably carbonic acid associated by exposure to what might seem to be insignificant concentrations of carbon dioxide.

Insight: A runaway response can develop for temperature control when a chemical reactor heat of reaction exceeds the cooling rate or for biomass concentration control when a bioreactor cell growth rate exceeds the death rate.

The primary process time constant for a process with positive feedback is the time for the PV to reach 172% of the value established by the open loop runaway process gain. This gain can be computed but roughly corresponds to the point where a noticeable continual acceleration becomes significant. The identification of the positive feedback time constant and open loop gain by an open loop test is rarely done in practice because of the danger of a runaway. These loops often have to stay in AUTO mode at all times so that the PID controller provides through its control algorithm the negative feedback missing in the process that is needed for safe operation.

Insight: Runaway processes are tuned as integrating processes. These processes are even more sensitive to the overuse of integral action. For safety reasons, some highly exothermic reactors have no integral action.

Runaway processes are identified as integrating processes. Even when a runaway process exists, tests are designed not to show the acceleration because such acceleration is unsafe. The tuning must be aggressive in that the process will run away if the PID gain is too small to provide enough proportional action. In the next section equations are presented to compute the maximum and minimum gain for stability as a function of dead time, the primary time constant and the secondary time constant. Proportional action is maximized to provide a more immediate reaction. Integral action is minimized (reset time is maximized) because the integral mode will not seek to add coolant until the PV has crossed setpoint leading to overshoot. Derivative action is more important than in other processes in order to help prevent excessive acceleration and to compensate for a secondary time constant.

Insight: Processes with positive feedback have a window of allowable PID gains where a PID gain too small and a gain too large will both cause a dangerous runaway condition.

Controller Tuning

There are over 400 of tuning rules compiled in Aidan O'Dwyer's *Handbook of PI and PID Controller Tuner Rules.* The author of each of these rules thinks their method is best. The diversity of tuning rules is primarily due to different types of process responses (e.g., dead time dominant self-regulating, moderate self-regulating, near-integrating, true integrating, and runaway processes), unknowns, nonlinearities, different objectives (e.g., setpoint response, load response, maximum variability absorption), and extenuating circumstances (e.g., interactions, inverse response, and resonance). The existence of runaway processes and the tuning guidelines have been directly addressed only by Bill Luyben in "Proportional-Derivative Control of Exothermic Reactors" and "Consider Reactor Control Lags". In the absence of additional guidelines, runaway processes should be treated as integrating processes. The use of derivative action is very important to help react to acceleration and provide anticipation.

If the effective rate time becomes larger than the reset time, oscillations can break out even if the tuning settings would otherwise provide stable operation. The Series PID Form used in analog controllers and DCS systems has inherently prevented the effective rate time from becoming more than the effective reset time by an interaction factor. Some tuning rules developed in the 1960s used rate times approximately equal to the reset time. The use of these settings can cause severe oscillations. There is no protection in the ISA Standard PID Form predominantly used today to prevent the rate time equaling or exceeding the reset time. To ensure the rate time is well below the reset time, a high limit for the rate time that is ¼ the reset time is shown in this book.

Insight: There is no protection in the ISA Standard Form predominantly used today to prevent the rate time equaling or exceeding the reset time causing severe oscillations.

The user must also be extremely careful about the units of the tuning settings and PID output scale. Some controllers use proportional band instead of gain for the proportional mode and repeats per minute instead of seconds for the integral mode. These settings have inverse relationships that can lead to enormous errors in tuning and performance. The output scale of the modern DCS is in engineering units even though the algorithm works with percent signals and the default output limits and anti-reset windup limits are in percent. The analog and early DCS PID controllers had output scales in percent.

Insight: In migration projects, the tuning settings and output and anti-reset windup limits must be carefully converted based on the PID Form, tuning setting units, and the PID output scale.

Most of the control literature focuses on improving the response to a setpoint or a disturbance on the process output for moderate self-regulating processes. Gamesmanship tends to rule and algorithms and tuning methods are touted based on improvements without a realization of the diversity of conditions and demands in actual industrial installations.

Lambda tuning rules are presented here because of their flexibility in dealing with different processes, nonlinearities, objectives, and extenuating circumstances.

Modifications are presented to the normal Lambda tuning rules to enable Lambda tuning to provide good load disturbance rejection capability when minimizing peak error and integrated absolute error is critical for largest spectrum of processes. For minimum errors from step load disturbances in studies, Lambda may be set equal to the dead time. In industrial applications, lambda is increased to provide robustness. Often a Lambda that is about 3 to 5 times the dead time is found to be more practical.

Lambda Tuning for Self-Regulating Processes

The Lambda tuning method for self-regulating processes is advocated by the author for use when the open loop time constant is less than 4 times the total loop dead time. For self-regulating processes that are not near-integrating ($\tau_o < 4 * \theta_o$) we have the following series of equations.

The reset time is set equal to the open loop time constant (Equation 1). The reset time steadily decreases as the ratio of the time constant to dead time decreases from its maximum of 4 dead times if this tuning method is used when the open loop time constant is less than 4 times the total loop dead time. For systems with an open loop time constant much less than the dead time (dead time dominant), the PID action becomes essentially integral only due to a steady decrease in the controller gain and the reset time. This type of control helps deal with the abrupt, almost step changes and noise in a dead time dominant system. Thus, the PID controller takes over the role of providing a gradual and smoothing response that is missing in the process. The best scenario is to have a large process time constant to fulfill this role but for plug flow and sheet line processes, this is not possible.

$$T_i = \tau_o \qquad (1)$$

A low limit is added to traditional equations to prevent a reset time smaller than ¼ the dead time when the time constant is much smaller than the dead time (Equation 1b). This low limit provides more proportional mode action (Equation 2a). Keeping the PID gain from approaching zero gives a more immediate response to disturbances and setpoint changes and helps get the PID output get through valve or VSD deadband and resolution limits.

$$T_i = Max\left[0.25 * \theta_o, \tau_o\right] \qquad (2)$$

Insight: For dead time dominant processes, a low limit of ½ the dead time is placed on the reset time to provide some proportional action

In industrial applications the Lambda factor multiplication of the open loop time constant is replaced with Lambda, the closed loop time constant ($\lambda = \lambda_f * \tau_o$), which is the time after the dead time to reach 63% of a setpoint change. The use of Lambda rather than a Lambda factor provides the recognition that Lambda is set relative to the dead time to minimize variability in the PID input or output and to provide the gain margin needed to deal with extenuating circumstances (e.g., interaction, inverse response, and resonance).

$$K_c = \frac{T_i}{K_o * (\lambda + \theta_o)} \tag{2a}$$

For maximum unmeasured disturbance rejection, a Lambda equal to the dead time is used giving Equation 2b for the PID gain setting when the reset time is not limited. For maximum disturbance rejection of self-regulating process, a Lambda equal to 0.6 times the dead time gives nearly the same PID gain as the original Ziegler Nichols reaction curve method defined in the landmark paper "Process Lags in Automatic Control Circuits". In practice, a Lambda greater than 3 dead times is used to deal with the uncertainties and nonlinearities in industrial processes.

$$K_c = 0.5 * \frac{\tau_o}{K_o * \theta_o} \tag{2b}$$

A rate setting is not normally included as part of the Lambda tuning for self-regulating processes. If the primary time constant is greater than ½ the dead time, rate action may be beneficial. If a secondary time constant (the next largest time constant) the rate time is set equal to the secondary time constant or½ the dead time, whichever is the largest, but not be greater than ¼ the reset time for an ISA Standard Form. The derivative mode must have a built-in filter whose time constant is a fraction of the rate time. The fraction is the alpha parameter in a PID block (e.g., 1/8).

If the primary time constant is greater than ½ the dead time ($\tau_o > 0.5 * \theta_o$):

$$T_d = Min\left[0.25 * T_i, Max\,(0.5 * \theta_o, \tau_s)\right] \tag{3}$$

Insight: When derivative action is beneficial the rate time should be the secondary time constant or ½ the dead time whichever is largest but not be greater than ¼ the reset time for an ISA Standard Form. The derivative mode must have a built-in filter whose time constant is a fraction of the rate time. The fraction is the alpha setting (e.g., 1/8).

Lambda Tuning for Integrating Processes

The Lambda tuning method for integrating processes is advocated by the author for use on self-regulating processes when the open loop time constant is greater than 4 times the total loop dead time (near-integrating processes) in addition to use on true integrating and runaway processes to provide a faster return to setpoint for load upsets. For near integrating processes, the open loop integrating process gain is the open loop self-regulating process gain divided by the open loop time constant. As a practical matter, this open loop integrating process gain is identified within four dead times as the maximum sustained % change in the ramp rate of the process variable in the correct direction divided by the % step change in the PID output.

The reset time is twice the arrest time plus the dead time.

$$T_i = 2 * \lambda + \theta_o \tag{4a}$$

A low limit is added so that the reset time is not less than 4 dead times. This limit only comes into play when a Lambda less than 150% of the dead time is used.

$$T_i = Max\left[4 * \theta_o, 2 * \lambda + \theta_o\right] \tag{4b}$$

In industrial applications the Lambda factor multiplication of the inverse of the integrating process gain is replaced with Lambda, the closed loop arrest time ($\lambda = \lambda_f / K_i$), which is the time to stop an excursion for an unmeasured disturbance (time to peak error). The use of Lambda rather than Lambda factor provides the recognition that Lambda is set as a multiple of the dead time to minimize variability in either the PID input or output and to provide the gain margin needed to deal with extenuating circumstances (e.g., interaction, inverse response, and resonance) or is set to provide the maximum absorption of variability for surge tank level control.

The numerator uses the unlimited reset time.

$$K_c = \frac{2 * \lambda + \theta_o}{K_i * (\lambda + \theta_o)^2} \tag{5a}$$

Insight: Near-integrating, true integrating, and runaway processes should use integrating process tuning rules with a low limit of 4 dead times for the reset time. The unlimited reset time is used in the numerator of the Lambda calculation for the PID gain.

For maximum unmeasured disturbance rejection, a Lambda equal to the dead time provides a controller gain that is about ¾ the gain from the Ziegler-Nichols reaction curve method:

$$K_c = 0.75 * \frac{1}{K_i * \theta_o} \tag{5b}$$

A Lambda approaching the dead time is only used for the highly unusual case where the dynamics are exactly known and all nonlinearities are exactly compensated for by adaptive control mostly. Low Lambdas are mostly used for gamesmanship to show how good your tuning method is at minimizing integrated absolute error. For example a Lambda of 5 dead times minimizes the consequences of nonlinearities, interactions, inverse response, and resonance.

Derivative action is particularly beneficial for integrating processes and necessary for runaway processes especially when there is an appreciable secondary time constant. In the Lambda tuning rules for integrating processes, derivative action is included with a rate time set equal to the secondary time constant. Here the rule is expanded to be the maximum of ½ the dead time and the secondary time constant to help deal with processes where multiple small lags create dead time. As with Equation 3, a high limit of ¼ the integral time (reset time) is used for the ISA Standard Form.

$$T_d = Min\left[0.25 * T_i, Max\,(0.5 * \theta_o, \tau_s)\right] \qquad (6)$$

Lambda tuning equations for integrating processes inherently protect against slow rolling oscillations from too low of a PID gain. In many industrial applications, the PID gain computed is much higher than the operations is comfortable with and will cause very large step changes of the PID output for setpoint changes unless moderated by a setup rate limits, setpoint filter, or PID structure. The reset time should be increased by the same factor that the PID gain is reduced as detailed in the next section on the Window of Allowable Controller Gains.

Insight: If less PID gain action is desired, Lambda should be increased rather than decrease to prevent slow rolling oscillations that can become nearly undamped.

Window of Allowable Controller Gains

The existence of a maximum PID gain is intuitive and extensively discussed or at least implied in control theory literature and courses. A PID gain just 50% higher than that computed with a Lambda equal to the dead time can cause the start of an oscillatory response with a period slightly larger than the ultimate period. Just a 50% increase in dead time or open loop gain can correspondingly trigger these same oscillations. An increase of 200% can cause oscillations to grow in amplitude with a period close to the ultimate period. Thus, the maximum controller gain for stability can be considered to be approximately three times the gain computed for a Lambda equal to the dead time (gain margin of 3) assuming all of the dynamics are exactly known and constant and there are no extenuating circumstances (e.g., interactions, inverse response, or resonance).

Rarely discussed in the literature is the existence of a low PID gain limit for near-integrating and true integrating processes when the integral mode is used and in runaway processes regardless of the modes used. In all of these cases, the loop depends upon gain action to provide negative feedback missing in the process dynamics.

Insight: For near-integrating, true integrating, and runaway processes, there is a window of allowable controller gains. Too high of a PID gain causes oscillations with a period close to the ultimate period and a growing amplitude (classic instability). Too low of a PID gain causes slow rolling oscillations with a period 10 times the ultimate period and a gradually decaying amplitude.

For near-integrating, true integrating, and runaway processes the minimum allowable PID gain to prevent slow rolling oscillations decreases as the reset time decreases and the open loop integrating process gain increases as detailed in Equation 7. The oscillations are extremely slow (e.g., 10 times the ultimate period). The oscillations start when the controller gain is less than 4 times the inverse of the product of the integral time and open loop integrating process gain ($K_x = 4$). The oscillations become nearly undamped when the controller gain is less than ¼ the inverse of the product of the integral time and open loop integrating process gain ($K_x = 1/4$).

$$K_{c\min} > K_x * \frac{1}{T_i * K_i} \tag{7}$$

Insight: For near-integrating, true integrating, and runaway processes the minimum allowable PID gain to prevent slow rolling oscillations decreases as the reset time decreases and open loop integrating process gain increases.

For runaway processes, the consequences of a low PID gain are more severe and potentially unsafe in that a reaction can go unstable from negative feedback action via the proportional mode being insufficient to counteract the acceleration from positive feedback in the process. The minimum PID gain per Equation 8 must be greater than the inverse of the open loop runaway process gain for temperature control of an exothermic reactor as defined in Figure 3.

$$K_{c\min} > \frac{1}{K_o'} \tag{8}$$

Insight: For runaway processes, an unsafe condition develops when the PID gain is less than the inverse of the open loop runaway process gain.

Luyben developed Equation 9 in "Proportional-Derivative Control of Exothermic Reactors" and "Consider Reactor Control Lags" to predict the ratio of the maximum to minimum controller gain (R_c) to show how the window narrows as the ratios increase of the heat transfer surface lag to positive feedback time constant (R_h) and of the temperature measurement time constant to positive feedback time constant (R_m). Luyben recommends a ratio greater than 20 to ensure that unknowns and nonlinearities or less than optimal tuning do not cause instability.

$$R_c = 1 + (1 - R_m - R_h) * \left[\frac{R_m + R_h}{R_m * R_h} - 1 \right] \tag{9}$$

Insight: For temperature control of exothermic reactors that can runaway, the window of allowable controller gains close as the sum of the ratios of the heat transfer and measurement time constant to the positive feedback time constant approach one.

Control Loop Performance

The integrated and peak errors for a step disturbance at the process input termed a load disturbance (e.g., feed flow or composition), can be estimated from equations developed from the reaction of the PID modes when the controller is in automatic. The integrated error is the area of deviation between the process variable and the setpoint on a trend chart until the process settles out a setpoint. The integrated error can be an inferential measurement of the accumulated amount of product that does not match specification (e.g., setpoint). If the response is non-oscillatory (over-damped), which is the desired case in industrial processes, the integrated error is the absolute integrated error (IAE), a metric commonly used in the literature to tout the value of an author's algorithm or tuning rule. The peak error is the maximum deviation of the process variable from setpoint for the load disturbance. The peak error is not commonly seen in the literature but is critical for preventing the triggering of undesirable and potentially unsafe side reactions and preventing high pressure and high temperature shutdowns.

A simple insightful equation for the integrated error (E_i) originally by Greg Shinskey is reformulated to express the disturbance as an open loop error, which is the error seen if the controller was in manual (loop was open). The resulting Equation 10 shows that the integrated error is proportional to the PID integral time (reset time) and inversely proportional to the PID gain. The use of Lambda equal to or slightly less than the dead time will minimize this metric. The value of this is mostly to reassure users that Lambda tuning rule of a PID can handle load disturbances about as well as the controller algorithms and tuning rules advocated in research papers. There is a tradeoff between performance and robustness (ability to maintain a non-oscillatory response despite uncertainties, nonlinearities, interactions, complex responses (e.g., inverse response), resonance, and slow secondary loops. The ability to deal with a wide spectrum of non-ideal and unknown conditions is a reality that often precludes tuning for minimum IAE. If the study does not show the trend charts, to ascertain a smooth response, the real value of the technology is unknown. For a perspective on this pervasive gamesmanship at play in the literature see the white paper "So Many Tuning Rules, So Little Time".

Equation 10 for the integrated error also shows that the size of the PID execution time denoted in DCS configurations as PID execution rate (Δt_x) and signal filter time (τ_f) relative to the integral time (reset time) determines the impact on loop performance.

$$E_i = \frac{T_i + \Delta t_x + \tau_f}{K_c * K_o} * E_o \tag{10}$$

Equation 11 for the peak error (E_x) is an extension to include the effect of the integral mode in an equation developed by Peter Harriott for a proportional-only controller. The effect of the integral mode on the peak error is usually only significant for balanced self-regulating processes. For dead time dominant processes, the peak error is basically the open loop error because the lack of a process time constant means the process variable change is so abrupt the controller does not have enough time to react.

$$E_x = \frac{1.25}{K_c * K_o * \left(1.0 + \frac{0.5 * \theta_o}{(T_i + \Delta t_x + \tau_f)}\right) + 1.0} * E_o \tag{11}$$

Temperature Measurement Performance and Dynamics

In industrial environments, high process temperatures, pressures, and vibration make it necessary to have a robust temperature sensor. Fast response time, accuracy, and stability are also needed. While several types of temperature sensors are available, such as thermistors, infrared pyrometers, fiber optic, and others, the two most commonly used in the process measurement industry are resistance temperature detectors (RTDs) and thermocouples (TCs).

RTD and TC Performance

The RTD provides sensitivity (minimum detectable change in temperature), repeatability, and drift that are an order of magnitude better than the thermocouple, as shown in Table 1. Sensitivity and repeatability are 2 of the 3 most important components of accuracy. The other most important component, resolution, is set by the transmitter. Drift is important for extending the time between calibrations and the temperature loop running at the right setpoint. The data in this table dates back to the 1970s and consequently doesn't include the improvements made in thermocouple sensing element technology and premium versus standard grades. However, the differences are so dramatic that the message is still the same.

Table 1 Accuracy, Range, and Size of Temperature Sensing Elements

Criteria	Thermocouple	Platinum RTD	Thermistor
Repeatability (°C)	1 - 8	0.02 - 0.5	0.1 - 1
Drift (°C)	1 - 20	0.01 - 0.1	0.01 - 0.1
Sensitivity (°C)	0.05	0.001	0.0001
Temperature Range (°C)	-200 - 2000	-200 - 850	100 - 300
Signal Output (volts)	0 - 0.06	1 - 6	1 - 3

The temperature range shown for the RTD in the table is optimistic. At temperatures above 500°C, changes in sensor sheath insulation resistance have caused errors of 10°C or more.

There are many stated advantages for thermocouples, but if you examine them more closely you realize they are not as important as perceived for industrial processes. Thermocouples are more rugged than RTDs. However, the use of good thermowell designs and good installation practices makes an RTD sturdy enough even for high-velocity streams and nuclear applications. Thermocouples appear to be less expensive until you start to include the cost of extension lead wire and the cost of additional process variability from less sensor sensitivity and repeatability.

The case of using TC or RTD input cards in a Distributed Control System (DCS) is not considered because of the error introduced by these input cards as a percent of span is large and individual sensor offset and drift errors cannot be individual corrected as they are with a dedicated temperature transmitter.

The IEC 751 standard describes an ideal relationship between the resistance of a platinum RTD and the temperature to which the RTD is subjected. The difference between the actual RTD curve and the ideal RTD curve results in a measurement error, which is referred to as *a sensor interchangeability error*.

The Callendar-Van Dusen equation offers an alternative to the IEC 751 standard. This equation can be programmed into a transmitter so that the transmitter can use the actual RTD curve rather than an ideal curve (e.g., IEC 751 standard) to translate the sensor's resistance signal into a temperature value. The Callendar-Van Dusen equation enables the transmitter calibration to be matched to the sensor. When a sensor is replaced the parameters for the equation are entered into the transmitter.

The installation of the sensor can introduce errors, noise, and dynamics causing poor measurement and control loop performance. Here we look at best practices to get the most out of the inherent capability of the sensor. The next post will provide guidance on the communication of the sensor signal to the control room to provide the best total installation.

Thermowell Length

To minimize conduction error (error from heat loss along the sensor sheath or thermowell wall from tip to flange or coupling), the immersion length should be at least 10 times the diameter of the thermowell or sensor sheath for a bare element. Thus, for a thermowell with a 1 inch (2.54 cm) outside diameter, the immersion length should be 10 inches (25.4 cm). For a bare element with a ¼ inch (6.35 mm) outside diameter sensor sheath, the immersion length should be at least 2.5 inches (63.5 mm). This is just a rule of thumb. Computer programs can compute the error and do a fatigue analysis for various immersion lengths and process conditions. For high velocity stream and bare element installations, it is important to do a fatigue analysis because the potential for failure from vibration increases with immersion length.

Thermowell Location

The process temperature will vary with process fluid location in a vessel or pipe due to imperfect mixing and wall effects. For highly viscous fluids such as polymers and melts flowing in pipes and extruders, the fluid temperature near the wall can be significantly different than at the centerline (e.g., 10 to 30°C). Often the pipelines for specialty polymers are less than 4 inches in diameter, presenting a problem forgetting sufficient immersion length and a centerline temperature measurement. The best way to get a representative centerline measurement is by inserting the thermowell in an elbow facing into the flow. If the thermowell is facing away from the flow, swirling and separation from the elbow as can create a noisier and less representative measurement. An angled insertion can increase the immersion length over a perpendicular insertion but the insertion lengths are too short for small diameter pipelines (e.g., pipe < 12 inches) unless the tip extends past the centerline. A swaged or stepped thermowell can reduce the immersion length requirement by reducing the diameter near the tip.

The distance of the thermowell in a pipeline from a heat exchanger, static mixer, or reactor outlet should be optimized to reduce the transportation delay but minimize noise from poor mixing or two phase flow. Generally 25 pipe diameters are sufficient to ensure adequate mixing from turbulence if there is a single phase, turbulent flow, and no great differences in the viscosity of streams being combined. Two phases exist for desuperheaters, split ranged transitions from cooling water to steam in jackets, the use of lime or ammonia as a reagent due to flashing and whenever slurries are involved.

The transportation delay will increase with distance adding more dead time to the loop. Consequently, there is a compromise between getting enough mixing to achieve a representative low noise measurement and creating too much additional dead time. In general, the transportation delay should be less than 10% of the PID reset time setting.

Insight: Generally a distance of 25 pipe diameters between the equipment outlet and the temperature sensor is sufficient to provide a relatively uniform temperature profile of a single phase fluid. The presence of different phases (e.g. bubbles or solids in liquids and droplets in steam) and high viscosity fluids will require longer distances.

Thermowell Construction

The stem of a thermowell is the part that is inserted into the process stream. Stems can be tapered, straight, or stepped. The performance of a thermowell varies with its stem design. In general, a tapered or stepped stem provides a faster response, creates less pressure drop, and is less susceptible to conduction error and vibration failure.

If the thicknesses of the thermowell walls and the fit of the sensing element are identical, thermowells with straight stems have the slowest time response because they possess the most material at the tip (largest diameter). Thermowells with stepped stems have the fastest time response because they possess the least material at the tip (smallest diameter). A small diameter also results in the least amount of drag force. Thermowells with stepped stems also provide the maximum separation between the wake frequency (vortex shedding) and the natural frequency (oscillation rate determined by the properties of the thermowell itself). If the wake frequency is 80% or more of the thermowell natural frequency, resonance and probably damage can occur. Generally, thermowells with tapered stems are slightly more expensive as a result of a more complicated manufacturing process.

Table 2 Effect of Process Velocity and Thermowell on Measurement Dynamics

Process Fluid Type	Fluid Velocity (feet per second)	Annular Clearance (inches)	Annular Fill Type	Time Constants (seconds)
Gas	5	0.04	Air	107 and 49
Gas	50	0.04	Air	93 and 14
Gas	150	0.04	Air	92 and 8
Gas	150	0.04	Oil	22 and 7
Gas	150	0.02	Air	52 and 9
Gas	150	0.005	Air	17 and 8
Liquid	0.01	0.01	Air	62 and 17
Liquid	0.1	0.01	Air	32 and 10
Liquid	1	0.01	Air	26 and 4
Liquid	10	0.01	Air	25 and 2
Liquid	10	0.01	Oil	7 and 2
Liquid	10	0.055	Air	228 and 1
Liquid	10	0.005	Air	4 and 1

Insight: Swaged, stepped, and tapered thermowells offer a faster response, lower pressure drop, and less possibility of vibration damage from resonance with wake frequencies.

The tip of the sensor must touch the bottom of the thermowell. Spring loaded sensor designs help ensure this is the case despite different installation practices and orientation. The fit of the sensor should be as tight as possible to reduce the annular clearance since air acts as insulator as seen in Table 2. The sensor lag can increase by an order of magnitude for a sloppy fit. For liquid systems, the additional lag effectively becomes an additional equivalent dead time in the measurement.

Insight: The tip of the temperature sensor must touch the bottom of the thermowell and the fit must be tight to prevent introducing a large sensor lag due to the low thermal conductivity of air.

Effect of Reactor Type on Dynamics and Control

Reactors involving liquid reactants can be categorized as continuous stirred tank reactors (CSTR), batch, fed-batch (semi-batch or semi continuous), and plug flow. In continuous reactors, there is a continuous discharge of product flow. A level controller is used in the CSTR to maintain a level that provides enough reaction time for a given production rate. In batch and fed-batch reactors, the discharge valve is closed until the batch is ready to transfer. A gas product or byproduct may be continuously generated during the batch and condensed and accumulated in an overhead system. A liquid and/or solids phase reaction without a continuous liquid and/or solids discharge flow is the distinguishing characteristic of batch and fed-batch. The implications as to dynamics, yield, and inventory control are significant.

Table 3 summarizes how the dynamic response, residence time, reaction rate and controlled variables vary with reactor type as discussed in the following sections. The reactor types considered are batch, fed-batch, plug flow, gas flow, and continuous stirred reactors (CSTR).

Table 3 Effect of Reactor Type on Dynamics and Control

(* - additional control besides temperature and pressure control)

Reactor Type	Dynamic Response	Residence Time Distribution	Reaction Rate	Additional Control *
CSTR	Near-Regulating Runaway	Wide	Moderate	Level Composition
Batch	Integrating Runaway	Very Tight	Slow	None
Fed-Batch	Integrating Runaway	Very Tight	Slow	Time Profile
Plug Flow	Self-Regulating Runaway	Tight	Fast	Length Profile
Gas	Self-Regulating Runaway	Tight	Very Fast	Composition

A continuous stirred reactor (CSTR) has a slow near-integrating response due to the residence time becoming a large primary time constant from good back mixing. A batch and fed-batch reactor has a slow true integrating response. A continuous plug flow reactor has a moderate self-regulating response for temperature control by manipulation of coolant due to the thermal time constant introduced by heat transfer surfaces. Inline polymerization reactors, static mixers, extruders, and gas reactors (e.g. fluidized bed reactors) are considered to be plug flow because there is negligible back mixing (negligible axial mixing). In gas reactors (e.g. fluidized bed reactors) feeds can be manipulated for temperature control because the reaction kinetics are so fast that there is little or no inverse response at the points chosen for temperature control. For these systems, coolant coils are switched in and out of service to set capacity. The process response is still moderate self-regulating when temperature manipulates feed due to the thermowell time constant. All of these reactors can develop a runaway response when the increase in reaction heat release with temperature exceeds cooling capability.

Insight: Temperature control of a continuous stirred reactor (CSTR) has a slow near-integrating response, a batch reactor has a slow true integrating response, and a continuous plug flow reactor has a fast moderate self-regulating response. All reactors can develop a runaway response if the exothermic reaction heat release rate exceeds the cooling rate capability.

The use of an analyzer with a large sample, cycle, and/or multiplex time almost guarantees dead time dominance (total loop dead time larger than open loop time constant) for composition control of continuous, plug flow, and gas reactors. Inferential measurements, online analyzers and fast at-line analyzers are used to minimize the introduction of excessive dead time into concentration control.

Insight: The use of slow at-line analyzers for concentration control of continuous reactors can create a dead time dominant loop.

The introduction of a slow temperature sensor or slow analyzer can be disastrous for reactors that can develop a runaway response. The dead time or secondary time constant must be much smaller than the primary positive feedback time constant otherwise the window of allowable controller gains closes causing instability for all tuning settings. Fouling of heat transfer surfaces for coolant and for thermowells add an excessive secondary time constant. Slow analyzers add excessive dead time.

Insight: The fouling of heat transfer surfaces and a long sample transportation cycle time or multiplex time can cause a reactor with runaway a response to be unstable for all tuning settings.

Sometimes fed-batch is called semi-batch or semi-continuous. However, there is an important distinction between fed-batch and continuous reactors; in that in fed-batch as in batch reactors there is no liquid product discharge flow until the end of the batch. The lack of a liquid discharge flow is an integrating response and a rise in level with reactant addition. The increase in level from the start to the finish of the batch and the lack of self-regulation as a result of the closed discharge valve affects the dynamic response of temperature and composition requiring tuning methods and settings not commonly discussed in the control literature. For example, a temperature controller gain of 50 or more is possible for fed-batch reactors.

Residence time is critical for reactors; it determines the amount of time available for reaction by virtue of the reactants being in contact with each other. The reaction time must be less than the residence time for the reaction to go to completion. If the residence time is too long, undesirable reactions and product degradation may occur. For maximum yield, the residence time should be slightly larger than the reaction time with a tight residence time distribution (statistical plot of each component population versus residence time should be a narrow peak). For maximum capacity and minimum capital cost, the residence time should be as small and the distribution as tight as possible since residence time is proportional to cycle time and volume.

Insight: For maximum process efficiency and capacity and minimum capital cost, the residence time should be slightly larger than reaction time with an extremely tight residence time distribution.

For batch operations the batch cycle time is extended past the feed phase via a hold phase to ensure the reaction is complete. The residence time is the average volume divided by average feed rate during the feed phase plus the time in the hold phase. For pure batch operations where all of the feeds are added quickly and simultaneously, all of the components are in the reactor for about the same time making the residence time distribution tight.

In fed-batch operations the feeds are added via a cascade control and ratio control system. Typically, a primary loop for concentration control is manipulating the setpoint of a secondary loop for the leader feed flow. The other feed flows have setpoints ratioed to the leader feed flow setpoint. The rate of feed addition for fed-batch operation is slower and nonlinear. The difference from one batch to another is not seen in the concentration profile but in the profile of the manipulated feed flows if the cascade control system is doing a good job. The residence time of components added early in the feed phase is longer adding a slight degree of uncertainty in the residence time distribution. Batch operations are considered here to be well-mixed so that turnover time is negligible. The downside of batch operations is a greater equipment cost to achieve the same capacity as continuous operations.

Insight: Batch operations assure a residence time greater than the reaction time and an extremely tight residence time distribution but offer less capacity than continuous operations.

The overall residence time for continuous operations is the current operating volume divided by the total flow rate passing through the volume. Some tuning methods for surge tank level control inadvertently use a definition of residence time as the maximum volume divided by the maximum change in flow set by equipment size, level connection location, and PID scale ranges. At high production rates, the residence time gets shorter for reactions unless the level setpoint is increased accordingly.

Insight: The residence time for continuous operations is the operating volume divided by the total flow.

A plug flow reactor has no back mixing (axial mixing). Polymer plug flow reactors use inline equipment such as pipes or tube(s). In static mixers baffles create some radial mixing. In extruders, a screw pushes a viscous liquid and solids to the exit. Since there is no back mixing all of the components in the reaction mass exit at the same time that is equal to the transportation delay. All of the residence time becomes a process dead time. A plug flow reactor is similar to a fed-batch reactor in that there is generally no level control and the residence time distribution is extremely tight. If you consider a subsection of the reaction mass moving from the entrance to the exit, there is no discharge from this volume until the subsection approaches the exit.

Insight: The residence time in plug flow reactors is a transportation delay and the residence time distribution is tight.

Polymerization processes often use batch and plug flow reactors operating with high viscosities. For highly viscous products the residence distribution is not as tight. Product can accumulate on the vessel walls in batch reactors. The velocity near the pipe wall in plug flow reactors is slower from frictional drag causing the product near the wall to arrive later. The increase in time available for reaction increases the viscosity of the fluid near the wall aggravating the problem.

Gas phase reactors are generally always continuous plug flow reactors. Many use fluidized catalyst beds. There is little to no back mixing except from turbulence. Consequently, the residence time distribution is tight for even catalyst distribution, the dynamic response to changes in reactant feed is a transportation delay similar to the plug flow liquid reactor except the reaction times are faster and the residence time shorter from a much higher velocity. Channeling of gas flows through the catalyst bed can result in some gases arriving at the outlet sooner reducing residence time and conversion for some portions of the exiting gas product.

Insight: The residence time of gas phase reactors without channeling is fast and tight.

In a continuous stirred reactor (CSTR) some of the reactants appear in the discharge within one turnover time, which in general is less than 1/20 of the residence time in a well-mixed vessel. The residence time becomes a large primary time constant causing a near integrating response. The amount of time spent by a given component in the reactor varies resulting in a looser residence time distribution and a loss in yield in the CSTR. Reactor designs with a liquid height about equal to the diameter, baffles to prevent swirling, axial agitation pattern that breaks the surface, and a dip tube design to bring reactants into the eye of the impeller; away from discharge nozzles to avoid short circuiting have a tighter residence distribution. Several CSTR in series or a reactant recovery and recycle system can be used to improve the yield of the production unit particularly for slow reactions and high purity requirements. However the use of a recycle system creates an integrating response and increases the propensity for buildup of inerts and impurities requiring the optimization of recycle flow and a purge flow for the best yield.

Insight: The residence time in continuous stirred reactors becomes a primary time constant and the residence time distribution is loose but can be tightened by good equipment design.

Liquid Reactor Control

Well mixed liquid reactors have dynamics that offer the possibility of extremely tight control of product composition but lack some of the self-regulation in multiple phase reactors achieved by inventory control. For a reactor to achieve its full potential requires an intelligent design of the control strategy. Here we look at how to setup up the control loops based with some guidance as to how the loops change for batch, fed-batch, or continuous reactors.

Basic Control

For reactions that are all in one phase (gas, liquid, or solid), inventory control (pressure or level control) cannot automatically adjust a reactant flow for changes in conversion (amount of product or byproduct formed per unit of reactant added) to prevent an excess or deficiency of reactant in a product phase (e.g., liquid) that is the opposite of the reactant phase (e.g., gas). Consequently, for single phase reactors, the use of analyzers is more important to control reactant stoichiometry for maximizing yield beyond what tight temperature control can do.

Insight: Single phase chemical reactors have a greater dependence on analyzers to maximize yield since there is not the inherent prevention of excess reactants by pressure or level control.

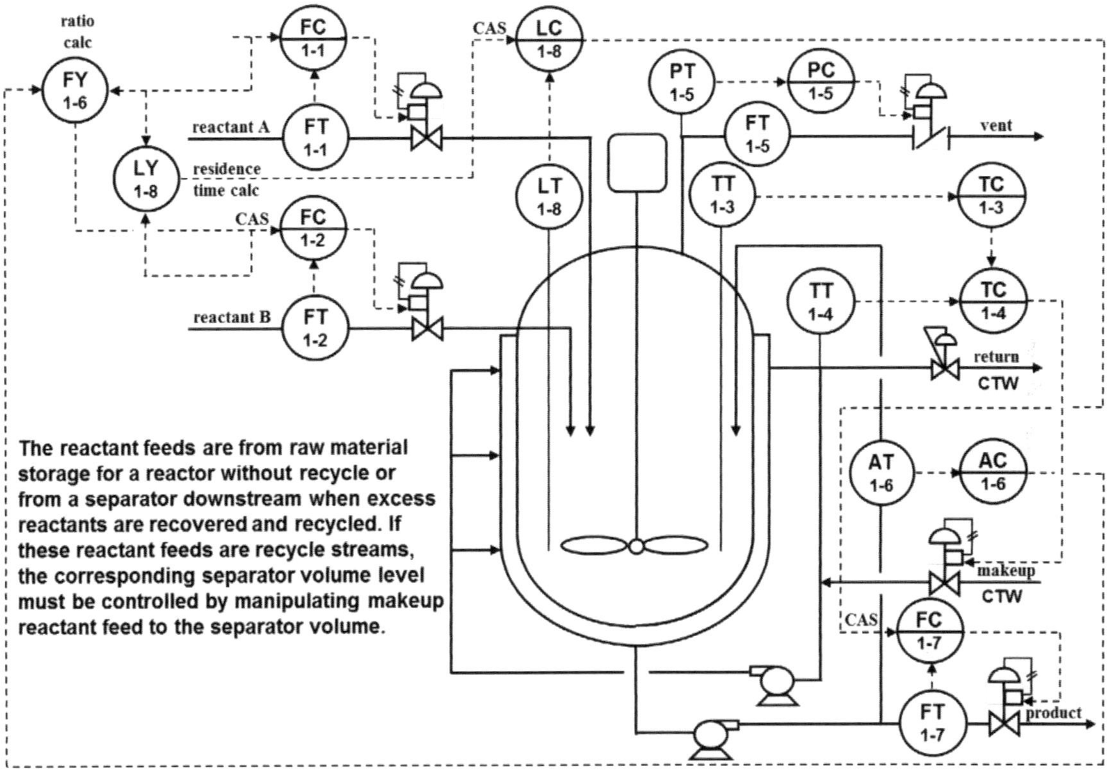

Figure 4 Liquid Reactor Basic Control

For well mixed single phase reactors the largest sources of improper reactant concentration leading to excess reactant are errors in reactant flow measurement and changes in reactant composition. Note that a deficiency in one reactant concentration creates an excess of another reactant. Coriolis meters can be used to provide the greatest mass flow measurement precision and rangeability with density correction for any changes in reactant feed concentration. Consequently, Coriolis meters on reactant feeds eliminate most of the sources of reactant unbalances if the mass flow ratios are correct and coordinated to maintain reaction stoichiometry.

Insight: Coriolis meters in reactant feeds are the best way to ensure correct stoichiometry if the feed setpoints are changed in unison with the correct ratio and the same closed loop response.

If the density of the excess reactant is significantly different than the density of the other components in the reactor, a Coriolis meter in the recirculation line can provide an inline inferential measurement of the excess reactant concentration to provide a fast and accurate correction of the ratio of reactants (Figure 4).

Insight: If changes in excess reactant concentration cause density changes greater than 1% and are at least 100 times larger than other density changes, a Coriolis meter in a recirculation line can be used as an inline analyzer for excess reactant concentration.

For continuous reactors with liquid reactants and liquid product, a level loop controls the inventory and thus the time available for reaction by manipulating the discharge product flow (Figure 4). If the desired reactor residence time multiplied by the total reactant flow is scaled and used as the setpoint of the level controller, the level loop helps maintain a constant residence time important to provide enough time for slow reactions. The reactor discharge flow is not used in the residence time calculation, to avoid the creation of a second loop within the level loop where a change in level PID output would cause a change in level PID setpoint which would cause a change in level PID output and so forth. The reactant feed setpoints are used instead of reactant flow measurements for the same reasons that setpoints are used in feedforward flow signals.

Batch reactors don't have a level loop and the concept of a residence time is replaced with a batch time for available time for reaction. The rising level of a batch reduces any self-regulation that might exist; changes in heat or mass transfer with temperature and concentration affect the open loop responses of temperature and product composition necessitating the use of higher controller gains to provide sufficient closed loop regulation. Integral action is also more likely to produce overshoot of setpoint.

Insight: A liquid level controller with a setpoint proportional to production rate can help provide the necessary residence time for liquid reactants in a continuous reactor.

Temperature loops control the energy balance and the reaction rate through the Arrhenius equation for chemical reaction kinetics. A cascade temperature control system offers the greatest linearity and responsiveness to coolant pressure and temperature upsets. The reactor temperature PID manipulates the setpoint of a jacket inlet temperature control that in turn manipulates makeup coolant flow. The coolant exit flow, such as cooling tower water (CTW) return flow, equals the coolant makeup flow to the jacket by piping design and in some case by pressure control in the recirculation system. The resulting constant jacket coolant flow eliminates the increase dead time, process gain and fouling from a decrease in jacket flow. An enhanced PID is used for the jacket inlet temperature loop to prevent limit cycling from valve stick-slip.

Insight: Temperature control determines the reaction rate via the Arrhenius equation.

Insight: Cascade control of reactor to jacket temperature isolates the reactor temperature PID from most of the nonlinearities and disturbances associated the jacket temperature response.

For batch reactors, as previously mentioned there is no level control and hence no residence time control but otherwise the temperature and gas pressure control systems shown for continuous liquid reactors are generally applicable. The tuning is even similar since a large primary process time constant enables continuous liquid reactor composition and temperature loops to be treated as having a near-integrating response that is similar to the integrating response of batch reactors.

Insight: The temperature and gas pressure control systems and PID tuning for batch reactors and liquid continuous reactors are similar.

For pure batch operations where flows are sequenced and scheduled based on batch times and totals the concentration endpoint is achieved by accurate charging of reactants and by a batch cycle time long enough to insure complete reaction. The simultaneous feeding of reactants by flow loops whose setpoints can be manipulated termed fed-batch control or semi-continuous control opens up opportunities for the use of concentration control as the batch progresses. The control schemes for concentration control of excess reactants in continuous liquid reactors, by correction of the reactant ratio shown in this chapter can be extended to fed-batch reactors. However, for non-reversible reactions the concentration control of product, as the batch progresses requires the translation of the controlled variable because the product concentration response is only in one direction. A translation to a rate of change of product concentration; slope of the batch product concentration profile provides a controlled variable that can increase and decrease based on feedback action. The computation provides an indication of product formation rate and yield and enables a prediction of batch endpoint.

Insight: Reactant concentration control systems for continuous liquid reactors can be extended to fed-batch reactors. However, product concentration control of non-reversible reactions requires a translation of the controlled variable to be the product concentration profile slope.

Analyzers and inferential measurements can provide composition control to correct the ratio of reactants. The lead reactant flow controller setpoint is multiplied by a ratio factor and added to the composition controller output. The setpoint is used instead of the actual flow measurement to reduce noise from the process and sensor (e.g. pressure and velocity profile fluctuations in differential head meters) and cycling from valve backlash or stick-slip and to enable better timing of the flow adjustments for production rate changes. However, the valve must be able to reach the flow setpoint by proper valve sizing and a valve design that prevents the valve from getting stuck or plugged. The reactant flow controllers are tuned for the same closed loop time constant to eliminate transient unbalances in the stoichiometric ratio of reactants added. A feedforward summer is used because both the process gain and the process time constant have an inverse relationship to reactant flow. Since the maximum controller gain is proportional to the ratio of process time constant to process gain, the effect of flow cancels out in the PID tuning.

At-line analyzers can be multiplexed with automated sample systems so one analyzer can service several reactors to help justify the cost. Inferential measurements can make the concentration measurement faster and smoother and less susceptible to missing or spurious results and noise. Inferential measurements except possibly for those from Coriolis meters must be periodically corrected by analyzer measurements ideally in the process but possibly in the lab. An enhanced PID is used in the composition loop to prevent excessive reset action and cycling from the large sample time and analyzer cycle time of at-line analyzers.

Insight: Analyzers and inferential measurements can provide concentration control to correct the ratio of reactants to account for unknowns in the reactants and reaction chemistry.

Insight: Inferential measurements can provide smoother and faster concentration measurements but do not eliminate the need for analyzers somewhere except for those measurements based on Coriolis meters.

Maximizing Production Rate

The control scheme commonly used by gas reactors that inherently maximizes production rate cannot be effectively used by liquid reactors. The optimization of production rate can be achieved by the simple configuration of a PID loop that uses existing signals. However, special PID features must be used to prevent oscillations and to deal with unmeasured disturbances.

The simplest control scheme for maximizing production rate in liquid reactors would be simply setting the coolant valve completely open and having the temperature controller bring in as much feed as the maximized coolant system can handle. However, the manipulation of reactant feeds by a temperature controller for maximization of production rate in liquid reactors causes inverse response for feeds colder than the reaction temperature and introduces a large lag that is the composition time constant of the vessel volume. For fast gas reactions, the residence time and consequently the composition time constant are small enough for this scheme to be effectively used, particularly when reactions are highly exothermic with a large heat generation that overwhelms any cooling effect of feeds.

Insight: The manipulation of reactant feeds for temperature control in liquid reactors does not work due to excessive inverse response and lag in the liquid composition response. The control scheme is productive for fast gas reactions.

In a first order plus dead time approximation for the temperature response, all time constants smaller than largest time constant are converted to equivalent dead time. The thermal process time constant for temperature control should be the largest time constant and the primary process time constant in the loop. The direct manipulation of feed rate for temperature control coupled with a slow reaction rate can cause a composition response time constant larger than the thermal time constant. The result is that a fraction of the thermal time constant becomes effectively dead time, causing a decrease in the maximum allowable PID gain and an increase in the minimum reset time and the peak and integrating errors for load disturbances.

Insight: A composition response in series with the thermal response in liquid reactors creates excessive dead time and deterioration in loop performance.

In a liquid reactor a coated temperature sensor or a sensor tip that does not extend past the nozzle into the vessel contents or a sensor hidden behind a baffle, could cause a measurement time constant larger that is than the thermal time constant. A sensor in a ceramic protection tube in a gas reaction phase or a sensor in a baffle with a glass coating in a small liquid volume will also have an excessive measurement lag due to the poor surface thermal conductivity of the sensor installation.

The slow composition response from manipulation of liquid reactant feeds causes slow correction and the slow sensor causes slow recognition of a disturbance. In either case, thermal time constants become effectively dead time and control severely deteriorates.

Insight: A sensor that is coated or is installed in a stagnant zone (e.g., behind a baffle) will add an excessive measurement lag.

The temperature sensor location is also important for the secondary jacket temperature loop. The sensor should be located in the coolant recirculation line rather than in the jacket. The higher velocities and turbulence in the pipeline provide a faster measurement with fewer fluctuations from level and phase changes and cold or hot spots from product sticking on the reactor wall.

Production rate can be maximized by the use of a valve position controller (VPC) monitoring coolant valve position (Figure 5). The VPC setpoint is the maximum desirable valve position, and the VPC process variable is the jacket temperature controller output. The use of actual valve position is unnecessary if the coolant valve has minimal backlash, good resolution (negligible stick-slip) and has a digital positioner. The maximum throttle position setpoint keeps the coolant valve near a point on the installed characteristic that has sufficient slope (valve gain) to correct for disturbances. The output of the VPC trims the setpoint of the "leader" reactant flow controller. An enhanced PID with external reset feedback (dynamic reset limiting) for the VPC eliminates limit cycles from coolant valve deadband, reduces interaction between the VPC and the jacket temperature controller, and enables smoother optimization with faster correction for large disturbances by directional move suppression. Directional move suppression by means of rate limits on the manipulated feed flow setpoints enables a gradual optimization and fast correction for abnormal conditions.

A reactant feed flow set point rate limit should be fast for correcting an increase in valve coolant position to prevent "running out of valve" for high heat releases. The setpoint rate limit should be slower for the opposite direction to provide a gradual optimization. External reset feedback in the PID automatically prevents integral action from driving the VPC PID output faster than rate limits allow or than the flow can respond. To see the increase in production rate in fed-batch operation from a higher feed rate, either the batch cycle time must be allowed to decrease or the batch mass allowed to increase.

Insight: A VPC, whose setpoint is the maximum coolant throttle position, can maximize feed rate smoothly with less interaction and better response to disturbances by the use of an enhanced PID and reactant feed setpoint rate limits to provide directional move suppression.

The reactant feeds could be fresh reactant streams from raw material storage or recycle streams when excess reactants are recovered in downstream separators. For the recycle case, the level in the corresponding downstream separator volume must be controlled by the manipulation of makeup reactant flow. For an excess reactant that is a heavier or lighter component (has a lower or higher boiling point) than the product, the separator volume that is the source of the reactant recycle flow, is a column sump or distillate receiver, respectively.

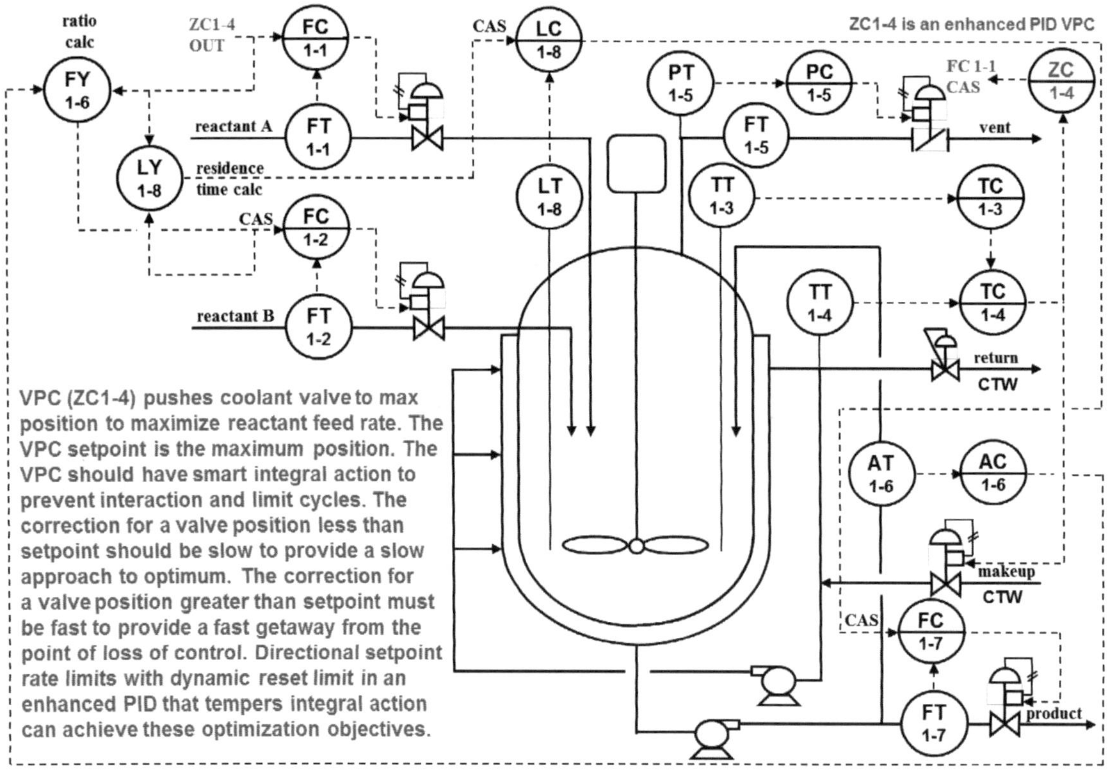

Figure 5 Maximizing Production Rate of Liquid Reactors

Prevention of "Snowballing Effect" from Recycle of Reactants

There must be a flow controller somewhere in the path of a recycle stream from reactor discharge to reactor feed to prevent the "snowballing effect", an escalating divergence of excess reactant concentration from any upset. The flow controllers on the reactant feeds serve this purpose when these are recycle streams.

Insight: When excess reactants are recovered and recycled to the reactor from downstream separators, a flow controller must exist in each recycle path to prevent the "snowballing effect".

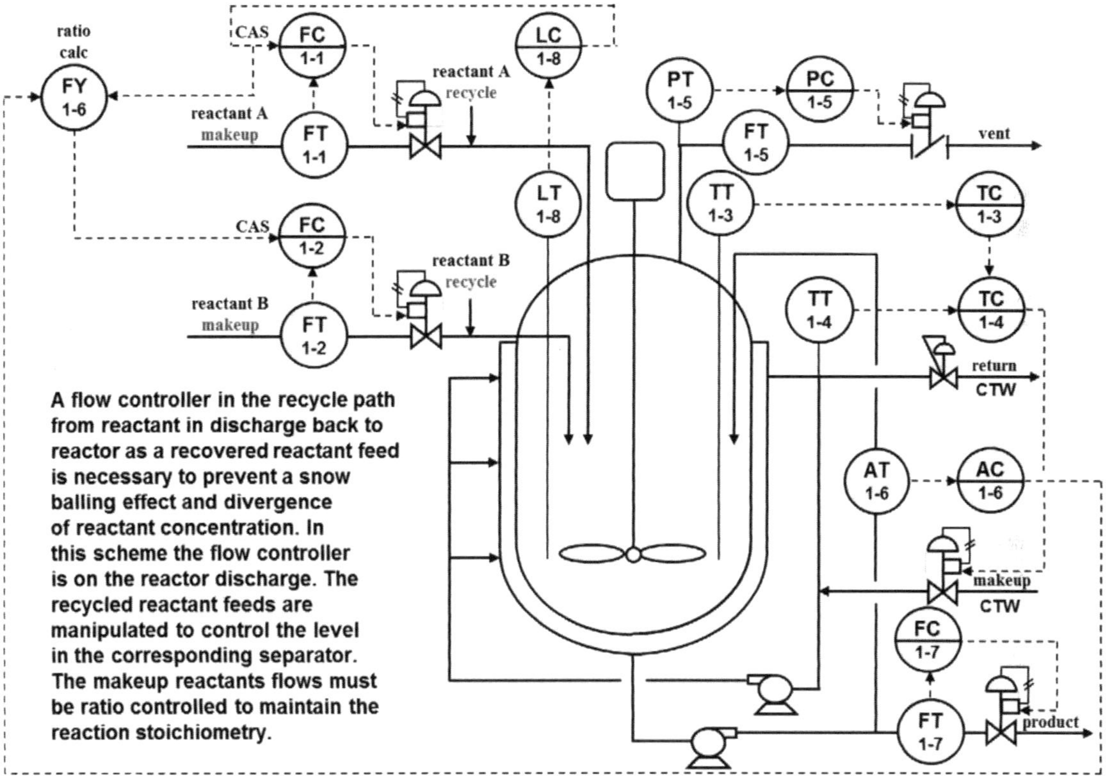

Figure 6 Prevention of "Snowballing Effect" of Reactant Recycle

An alternate control scheme for recycle systems is used when the reactor level controller manipulates a makeup reactant flow (Figure 6). The reactor discharge flow controller sets production rate and prevents the snowballing effect. The makeup reactant feeds must be ratio controlled to maintain stoichiometry. If reactant B flow is left on local flow control, the reactor concentration of reactant B will go low for an increase in production rate or a disturbance that caused a decrease in in reactor level. An accurate analyzer is critical in this control scheme to correct the ratio because the lack of inherent self-regulation can cause a continual error in the stoichiometry.

Insight: If the makeup reactant flows are set by the reactor control system rather than corresponding downstream separator volume level controller, the makeup reactant flows must be accurately ratio controlled to maintain stoichiometry for changes in level.

Fluidized Bed Reactor Control

For gas reactants and a gas product, a pressure loop controls the material balance and provides the time available for reaction at a given production rate by manipulating the discharge product flow to balance the total feed flow (Figure 7). The residence time for these fast reactions is small (e.g., a few seconds) but still must be kept above a low limit. A fluidized catalyst bed is used to promote reaction rate. If a temperature loop manipulates the leader gas reactant flow, the production rate is automatically maximized by the temperature and pressure controllers for a given cooling rate established by BFW flow and the number of coils in service as discussed below. Valve position control is not needed unless a control strategy is added to manipulate the BFW. Direct manipulation of feed rate by the temperature control is possible in gas reactors because the additional time lag for composition response is negligible due to the small residence time and the inverse response at the control points is negligible due to the fast reaction, high heat release, and catalyst heat capacity.

Insight: Temperature control of fluidized bed gas reactor by manipulation of feed rates will inherently maximize feed rate for a given cooling capability with no appreciable inverse response or composition response lag.

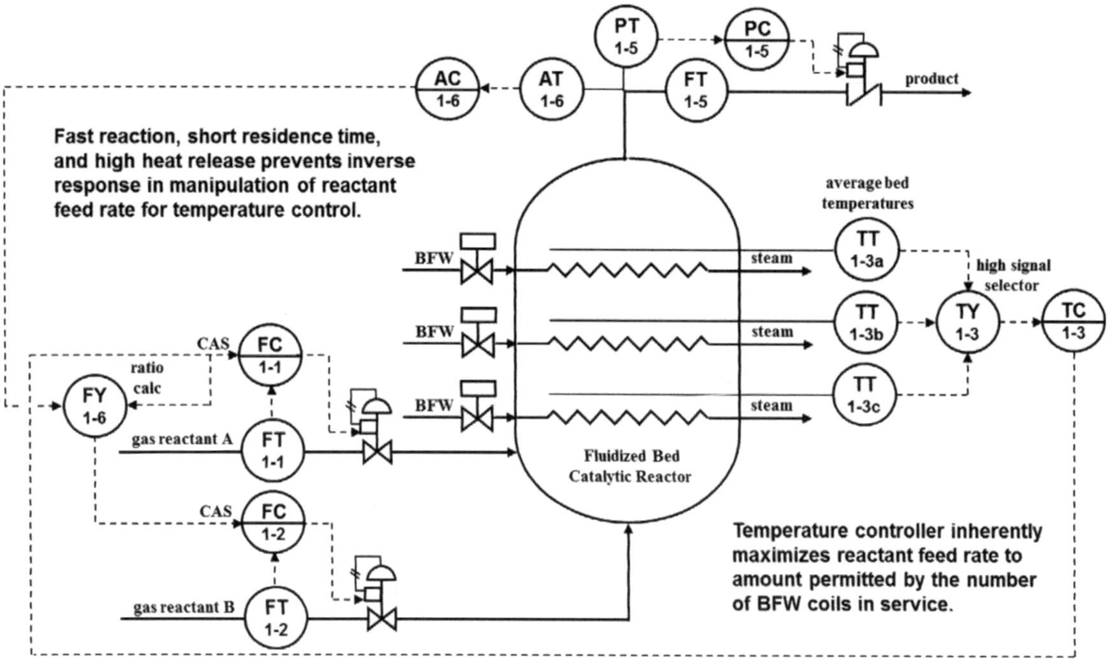

Figure 7 Fluidized Bed Reactor Control

A gas reactor with a fluidized catalyst bed may develop hot spots from localized high reactant concentrations due to a non-uniform flow distribution and no back mixing. Numerous separate cooling coils are used so operations personnel can switch coolant coils in or out of service to deal with hot spots and changes in production rate. However, the switching causes a disturbance to the temperature controller as fast as the BFW on-off valves can move. Numerous thermowells each with multiple sensors traverse the reactor. The average temperature is computed for each traverse with the highest average selected as the control temperature. Only 3 thermowells and BFW coils are shown in Figure 7 due to pictorial space limitations. Special designs can maximize the contact between the sensor tip and the thermowell wall to minimize the sensor lag. Since the gas velocity is usually quite high, the sensor time constants can be small if the insulating effect of air in the thermowell at the sensor tip is minimized. Note that this sensor time constant is the largest time constant in the loop and can mislead one into thinking a slower measurement is better due to the filtering effect and the larger maximum allowable temperature controller gain as detailed in the 12/02/14 Control Talk Blog "Measurement Attenuation and Deception Tips"

A feedforward signal can provide preemptive correction for the disruption of coil switching by means of a gain and velocity limit set to match the BFW on-off valve installed characteristic slope and stroking time.

Insight: Fluidized bed reactors use the highest of the average bed temperature at various distances in the flow direction to control hot spots.

Insight: A feedforward signal based on the installed characteristic and stroking time of the BFW on-off valves can be used to reduce the temperature upset from the switching of coolant coils.

The plug flow of reactants through the reactor provides a tight residence time distribution. To provide a residence time greater than the reaction time for the greatest production rate, the gas volume must be large enough and the flow distribution uniform enough for the reaction to go to completion.

Changes in discharge composition are mainly due to errors in the flow measurement, hot spots triggering side reactions, or insufficient radial mixing. An excess of one reactant is often used to ensure complete conversion of other reactants. An at-line analyzer on the gas product can be used to correct the gas reactant ratio to improve yield by reducing excess reactant.

The leader reactant flow multiplied by a ratio factor is the feedforward signal for the composition controller. A feedforward summer is used even though a feedforward multiplier can compensate for the inverse relationship between process gain and flow because of scaling and analyzer reliability issues and the predominant error seen is a bias rather than a span error in the flow measurements. The composition loop trims the feedforward signal. An enhanced PID with a threshold sensitivity setting helps deal with the analyzer sample and cycle time and the noise from poor mixing.

Insight: Plug flow reactors have a tight residence time distribution but no opportunity for residence time control by level control. The reactor volume must be large enough and the flow distribution uniform enough to provide enough residence time at the highest production rate.

Insight: An at-line analyzer and inferential measurement can reduce the excess concentration of a reactant used to ensure the complete conversion of other reactants.

Mixed Phase Chemical Reactor Control

Shinskey offers key guidance on reaction completion control in his ISA book *Controlling Multivariable Processes* that is conceptualized and extended here to include solid phases and batch operations. The term "completion control" is employed instead of the term "endpoint control" used by Shinskey because the term "endpoint" outside of the oil, gas, and petrochemical industry is considered to designate the conditions at the end of a batch operation.

For both batch and continuous reactors, completion control seeks to provide a complete conversion of all the reactants and consequently no excess accumulation of a reactant in a particular phase. If the reactants are in different phases and the product is a single phase, inventory control can be used for reaction completion control. The product must be a gas, liquid, or solids with no recycle or co-products in the other phases.

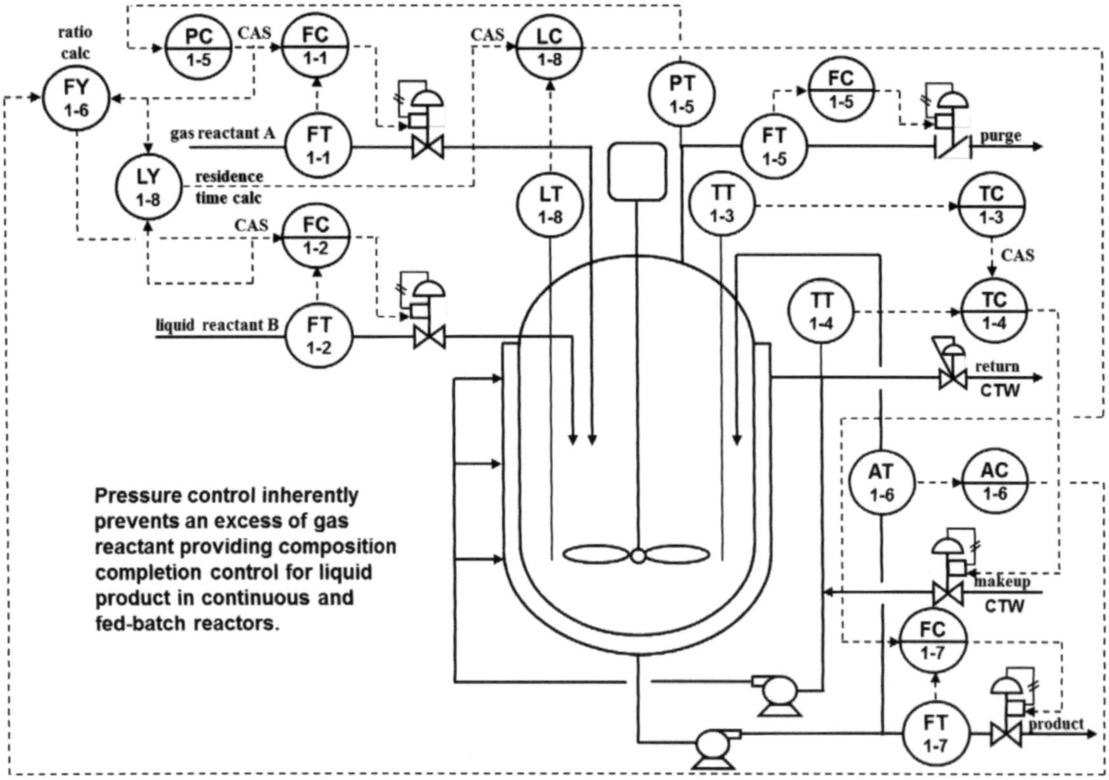

Figure 8 Basic Control of Multiphase Reactor with Liquid Product

Inventory control is possible if there is a discharge flow in addition to a feed flow to the phase. Gas volumes have a continuous discharge flow in batch as well as in continuous reactors by means of an off-gas flow going to a vent or recovery system. Liquid and solid volumes only have a discharge flow in continuous operations unless there is a phase separator physically built-in to continuously separate solids from liquids during batch operation (e.g., a hydroclone in a recirculation line).

Insight: Completion control seeks to provide for batch and continuous reactors a complete conversion of all reactants.

Using these concepts one can determine whether inventory control provides completion control. For a liquid product, excess gas reactant is inherently prevented by pressure control for both batch and continuous processes. For a gas product, excess liquid reactant is inherently prevented by level control for continuous processes and possibly solids control for all processes.

Insight: For multiple phase reactants and a single phase product, inventory control can provide reaction completion control. Batch processes do not have liquid inventory control (level control) to prevent the accumulation of reactants in the liquid phase.

For a liquid product, a pressure controller provides continuous completion control by increasing gas feed for a decrease in pressure from a deficiency of gas reactant and by decreasing gas feed for an increase in pressure from an excess of gas reactant (Figure 8). The gas phase reaction is normally fast enough for the gas reactant to be totally consumed in the reaction so that only the inerts are left in the overhead vapor space. An off-gas purge flow prevents the accumulation of inerts. A level controller maintains the liquid material balance by manipulating the liquid product discharge flow.

The residence time control by the level loop shown for a single phase liquid reaction (Figure 4) could help maintain the residence time in the gas phase as well the liquid phase by keeping the bubble rise time constant. The bubble rise time is the bubble path length set by level divided by the bubble velocity set by sparge rate. For pure batch and fed-batch reactors there is no level control and hence no residence time control but otherwise the control scheme is applicable.

Insight: An off-gas purge of inerts enables gas pressure control to provide reaction completion control in the gas phase for a liquid product.

For a gas product, a level controller provides continuous completion control by increasing liquid feed for a decrease in level for a deficiency of liquid reactant and decreasing liquid feed for an increase in level from an excess of liquid reactant (Figure 9). A purge flow from the bottom prevents the accumulation of inerts in the liquid phase. A gas pressure controller maintains the gas material balance by manipulating the gas product discharge flow. For residence time control, the liquid controller setpoint computed as the liquid reactant flow multiplied by the desired residence time would need a lag inserted because the level controller manipulating the liquid reactant flow forms a positive feedback loop. For fed-batch reactors this strategy is not applicable because there is no level control or purge.

For a gas product, a level controller provides continuous completion control by increasing liquid feed for a decrease in level for a deficiency of liquid reactant and decreasing liquid feed for an increase in level from an excess of liquid reactant (Figure 9). A purge flow from the bottom prevents the accumulation of inerts in the liquid phase. A gas pressure controller maintains the gas material balance by manipulating the gas product discharge flow. For residence time control, the liquid controller setpoint computed as the liquid reactant flow multiplied by the desired residence time would need a lag inserted because the level controller manipulating the liquid reactant flow forms a positive feedback loop. For fed-batch reactors this strategy is not applicable because there is no level control or purge.

Insight: A bottom liquid purge of inerts enables liquid level control to provide reaction completion control in the liquid phase for a gas product.

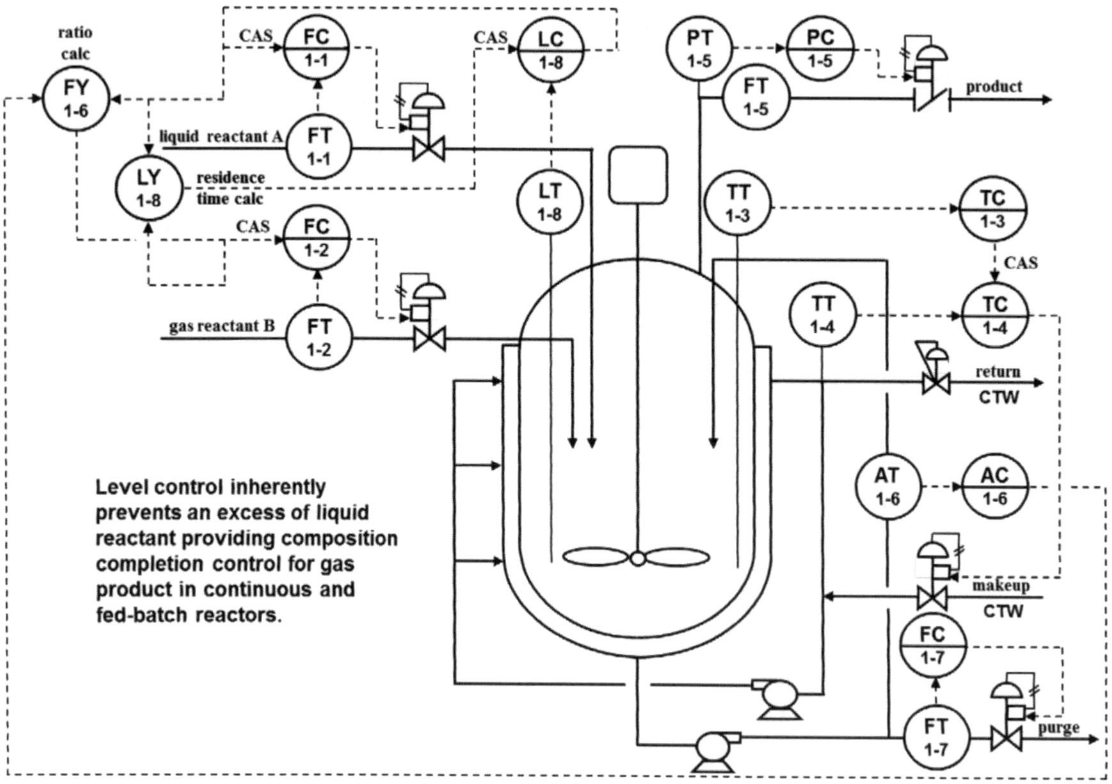

Figure 9 Basic Control of Multiphase Reactor with Gas Product

The purge flow for both types of reactor presents an opportunity for optimization. An excess accumulation of inerts reduces the residence time and reaction rate. Inhibiters can reduce yield and initiators of undesirable compounds can cause poor product quality besides reducing yield. First principle models can provide an inferential measurement of inert, inhibiter, or side reaction initiator accumulation. The predicted accumulation is corrected by at-line or off-line analyzers of the concentrations in raw materials (reactants) and phases within the reactor. It is important that research studies be conducted on the type of inerts, inhibiters, and initiators of undesirable reactions that can be found in the reactant feeds and their effect on reaction rate and yield.

Production rate can be maximized in both cases by the use of a valve position controller (VPC) monitoring coolant valve position. The VPC setpoint is the maximum desirable valve position, and the VPC process variable is the temperature controller output. The output of the VPC trims the setpoint of the liquid and gas reactant flow controller for liquid and gas products, respectively. An enhanced PID with dynamic reset limiting for the VPC eliminates limit cycles, reduces interaction between the VPC and temperature controller, and enables smoother optimization with faster correction for disturbances.

If there are products in different phases, continuous completion control requires an online or at-line analyzer because the level and pressure controllers have to regulate the liquid and gas, respectively, for inventory control. Analyzers can provide composition control by manipulating the ratio of reactants or a two phase product flow.

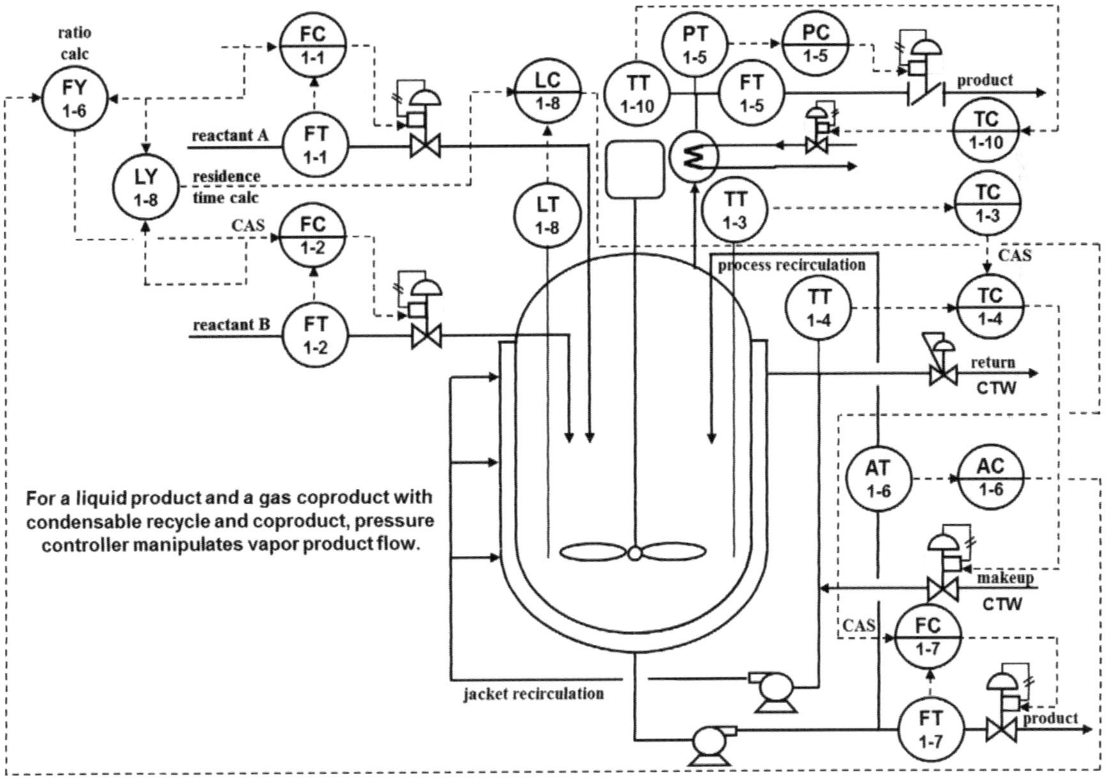

Figure 10 Basic Control of Multiphase Reactor with Liquid and Gas Coproducts

For a liquid product and a vapor co-product with condensable recycle, a reactor pressure controller manipulates vapor co-product flow. A reactor vapor temperature controller manipulates partial condenser coolant flow and a reactor liquid temperature controller manipulates the secondary jacket coolant temperature setpoint (Figure 10). The reactor level controller manipulates liquid product flow. The level setpoint can be computed to provide residence time control. For fed-batch reactors there is no reactor level control and hence no residence time control but otherwise the same control scheme is applicable.

Insight: For liquid and vapor co-product with recycle from a vent condenser, a gas pressure control manipulates vapor flow to a recovery system, a condenser vapor temperature controller manipulates condenser coolant flow, and a reactor liquid temperature controller manipulates jacket temperature.

Production rate can be maximized by the use of a valve position controller (VPC) monitoring condenser and jacket coolant valve positions (Figure 11). The VPC setpoint is the maximum desirable valve position, and the VPC process variable is the highest of the reactor vapor and liquid temperature controller outputs. The output of the VPC trims the setpoint of the "leader" reactant flow controller. An enhanced PID with dynamic reset limiting for the VPC eliminates limit cycles, reduces interaction between the VPC and the temperature controllers, and enables smoother optimization with faster correction for large disturbances.

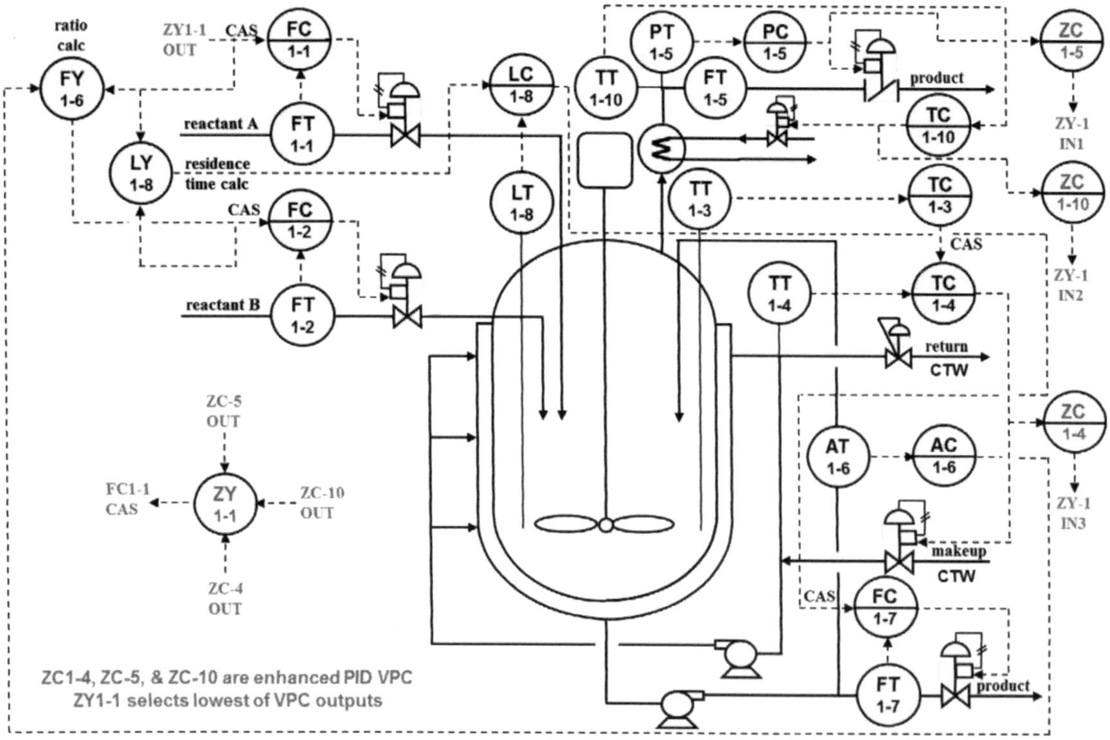

Figure 11 Maximizing Production Rate of Multiphase Reactor with Liquid and Gas Coproducts

If there is no vapor product or vapor exit flow, there is a vapor-liquid equilibrium relationship where the gas pressure determines the liquid temperature and vice versa. To eliminate the temperature measurement lag and mixing non-uniformities, an inferential measurement of temperature can be computed from gas pressure using a simple straight line approximation to the pressure versus temperature relationship.

The inferential temperature controller called a P-T controller manipulates the jacket temperature. The P-T controller offers a significantly faster and more stable loop response than a temperature controller particularly for long immersion thermowells, coated thermowells, and a low mixing velocity. For glass lined vessels where the thermowell has a glass exterior, the thermal lag eliminated is huge. For highly exothermic reactors with potential a runaway response, a large measurement lag reduces the ratio of the maximum to minimum controller gain for stability as detailed in Chapter 1 making the reactor more susceptible to nonlinearities. An integral-only temperature controller is used to gradually correct the inferential measurement of the P-T controller by the addition or subtraction of a simple bias. This scheme has been successfully applied to large batch polymerization reactors.

Insight: If there is no vapor product or vapor exit flow, an inferential temperature measurement computed from pressure based on vapor-liquid equilibrium can be used to eliminate the temperature sensor lag and mixing non-uniformities to provide a faster and smoother temperature control.

Acid and Base Chemical Reactor Control

For acid and base reactions, pH can be an extremely sensitive indicator of excess acid or base offering continuous completion control of both reactants. The same strategy used for liquid reactants and liquid product is applicable with the analyzer AT1-6 being pH (Figure 4). Control valves with a resolution better than 0.2%, such as diaphragm actuated sliding stem valves, are more generally available for smaller flows to provide more precise reactant manipulation. In these cases, the largest reactant flow is chosen to be the "leader" leaving the smaller flow for pH control.

Consider the largest flow being an acid (reactant A) whose flow controller sets reaction rate. The base (reactant B) flow multiplied by a ratio factor becomes the feedforward signal for a pH controller manipulating the acid reactant flow setpoint. The pH controller trims the feedforward correcting for bias errors in the flow measurements.

Insight: For acid and base chemical reactors, pH can be a sensitive indicator of excess acid or base concentration and can provide precise reaction completion control.

A reactor level controller manipulates discharge product flow to control inventory and residence time. For fed-batch reactors there is no reactor level control and hence no residence time control but otherwise the same control scheme is applicable.

Production rate can be maximized by the use of a valve position controller (VPC) monitoring acid or base reactant valve positions just as depicted for the liquid chemical reactor (Figure 4). The VPC setpoint is the maximum desirable valve position, and the VPC process variable is the temperature controller output. The output of the VPC trims the setpoint of the leader reactant flow controller. An enhanced PID with dynamic reset limiting for the VPC eliminates limit cycles, reduces interaction between the VPC and the temperature controller, and enables smoother optimization with faster correction for disturbances.

Mammalian Bioreactor Control

Mammalian bioreactors have the most stringent control requirements particularly for pH and temperature. The liquid termed media consists of water, nutrients, and a substrate of sugar (glucose) and amino acid (glutamine). Agitation is kept at a low rate because mammalian cells have no cell walls (they have membrane) and are easily damaged by impeller shear. Mammalian bioreactors have all of the process control for bacterial bioreactors but with more complex DO and pH loops (Figure 12). Sparger design is critical to provide dispersion of gases and mixing and reduce interaction between the DO and pH controllers.

In an effort to increase the safety, efficiency and affordability of medicines, the Food and Drug Administration (FDA) has proposed Process Analytical Technology (PAT) guidance to encourage innovation and improvement in pharmaceutical development, manufacturing, and quality assurance. The primary focus of the PAT guidance is to reduce variability through a better understanding of a process than can be obtained by the traditional approach. PAT is formally defined by the Food and Drug Administration (FDA) in Part IV of the PAT guidance as follows:

"The Agency considers PAT to be a system for designing, analyzing, and controlling manufacturing through timely measurements (i.e., during processing) of critical quality and performance attributes of raw and in-process materials and processes, with the goal of ensuring final product quality. It is important to note that the term analytical in PAT is viewed broadly to include chemical, physical, microbiological, mathematical, and risk analysis conducted in an integrated manner. The goal of PAT is to enhance understanding and control the manufacturing process, which is consistent with our current drug quality system: quality cannot be tested into products; it should be built-in or should be by design."

While analyzers are the key to better process understanding and control, PAT is more than just about adding analyzers. PAT seeks to promote better use of the data. The term "analytical" refers the ultimate goal of developing knowledge of the nature, proportion, function, and interrelationship between the constituents of the manufacturing process.

Insight: Mammalian cell bioreactors have more stringent control requirements and a greater opportunity for PAT initiatives and process optimization due to the sensitivity of the cells to operating conditions and sophistication of the cell functions.

Basic Control

The DO controller output is split ranged to manipulate air and oxygen sparge and air overlay (vapor space addition) as shown in Figure 12. The pH controller output is split ranged between carbon dioxide addition (carbonic acid) and sodium carbonate (base). These are weak acids and bases, providing a great moderating (buffering) effect on the titration curve slope and enabling much tighter pH control.

Weak acids and bases, and the use of a tiny aperture for the reference electrode junction, reduce the liquid junction potential providing a more accurate pH measurement. The disturbances from changes in cell concentration and function take days to develop (slower than concentration disturbances in any other type of reactor). With proper tuning, pH control within 0.02 pH is possible. Instead of dealing with cell disturbances, the focus is on eliminating overshoot for changes in pH setpoint, reducing interaction with the DO control, and preventing high carbon dioxide concentrations that can suffocate cells and the excessive accumulation of sodium carbonate concentrations that can increase the osmotic pressure (cell internal pressure).

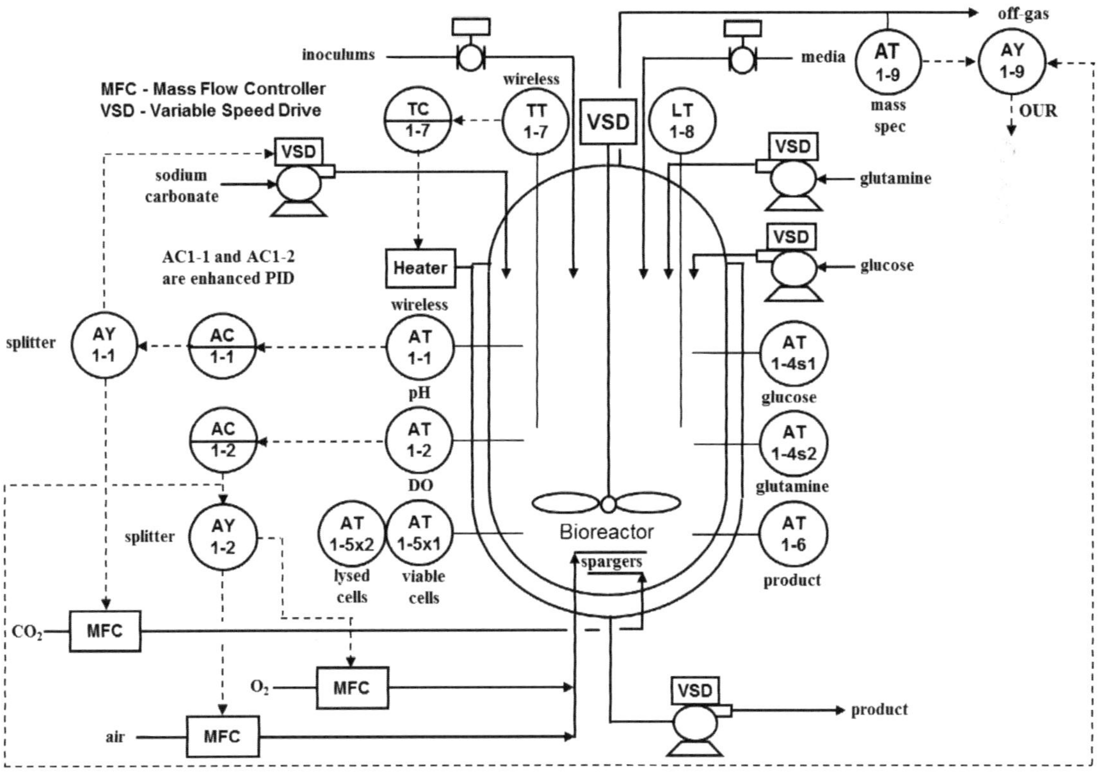

Figure 12 Basic Control of Mammalian Reactor

Insight: Incredibly tight pH control is possible because of buffering and extremely slow disturbances.

Interaction between DO and pH control can be greatly reduced by sparger design. Remaining interaction can be dealt with by an enhanced PID for DO and pH control with a threshold sensitivity setting for ignoring noise and a decoupling feedforward signal that is the opposing controller's gas addition rate. Since tight pH control is more important than DO control, a feedforward of DO PID output (e.g., air flow setpoint) can be added to the pH PID output to help the pH controller immediately correct the carbon dioxide (CO_2) flow setpoint.

The feedforward does not need to correct the pH PID output when manipulating the sodium carbonate setpoint. This simple half decoupler of air to CO_2 flow setpoint is usually sufficient unless there is a sparger design problem. Problems occur when air and CO_2 flow share the same sparger, the spargers are too close to each other, or there is no tight pressure control of air or CO_2.

Insight: Good sparger design can eliminate most of the interaction between pH and DO control. What remains can be reduced by a half decoupler where the air flow setpoint from the DO PID output is added to the pH PID output that is the CO_2 flow setpoint.

High concentrations of carbon dioxide (CO_2) can be avoided by preventing overreaction of the pH controller to pH shifts (pH setpoint changes to promote product formation) by the use an enhanced PID and a secondary CO_2 flow PID with setpoint rate limits. High CO_2 concentrations will dissipate with time. Inert gases are sometimes added to help sweep CO_2 out of the batch.

Mammalian bioreactor batches take 10 or more days. The exceptionally slow kinetics translate to extremely slow changes in reagent, oxygen, and cooling demand. Exceptionally slow load disturbances and slow integrating process gains result in negligible errors from disturbances. The objectives of tuning and control loop performance for pH, dissolved oxygen, and temperature control are in terms of setpoint response and optimization rather than disturbance rejection. The recognized opportunity is getting to a setpoint with no overshoot for a temperature shift and pH shift (setpoint change) to move from enhancing cell growth to promoting product formation. A PID structure that has proportional action on the process variable rather than on error and no derivative is useful for preventing overshoot. Getting to setpoint faster is not a consideration since the minutes saved are inconsequential in terms of the batch cycle time.

Insight: Mammalian cell kinetics are extremely slow with batch cycle times of 10 or more days. The errors from load disturbances are so slow that tuning for disturbance rejection or getting to setpoint faster is not the goal. Most of the focus is on the prevention of overshoot.

As previously mentioned, excessive accumulation of sodium carbonate causes a high sodium concentration leading to a high osmotic pressure that can contribute to cell membrane rupture and cell death. Unlike carbon dioxide that can escape in the vent gas, sodium carbonate is trapped in the media. Excessive crossing back and forth of the split range point for pH control over the course of the batch can lead to high osmotic pressure. A split range gap or deadband complicates tuning and does not stop the eventual excursion from the integrating process response. External reset feedback (dynamic reset limit) in the primary pH PID and setpoint rate limits in the secondary flow PID can provide the directional move suppression needed to slow down a movement into the direction of adding sodium carbonate reducing unnecessary split range excursions without requiring special tuning. New at-line analyzers can measure osmolality (osmotic pressure) by freeze point depression to help in optimization of the setpoint rate limits.

Insight: External reset feedback in the pH PID and setpoint rate limits in the CO_2 and the sodium carbonate flow controllers can provide directional move suppression to eliminate unnecessary crossings of the split range point and an increase in osmotic pressure.

Glutamine to Glucose Ratio Control

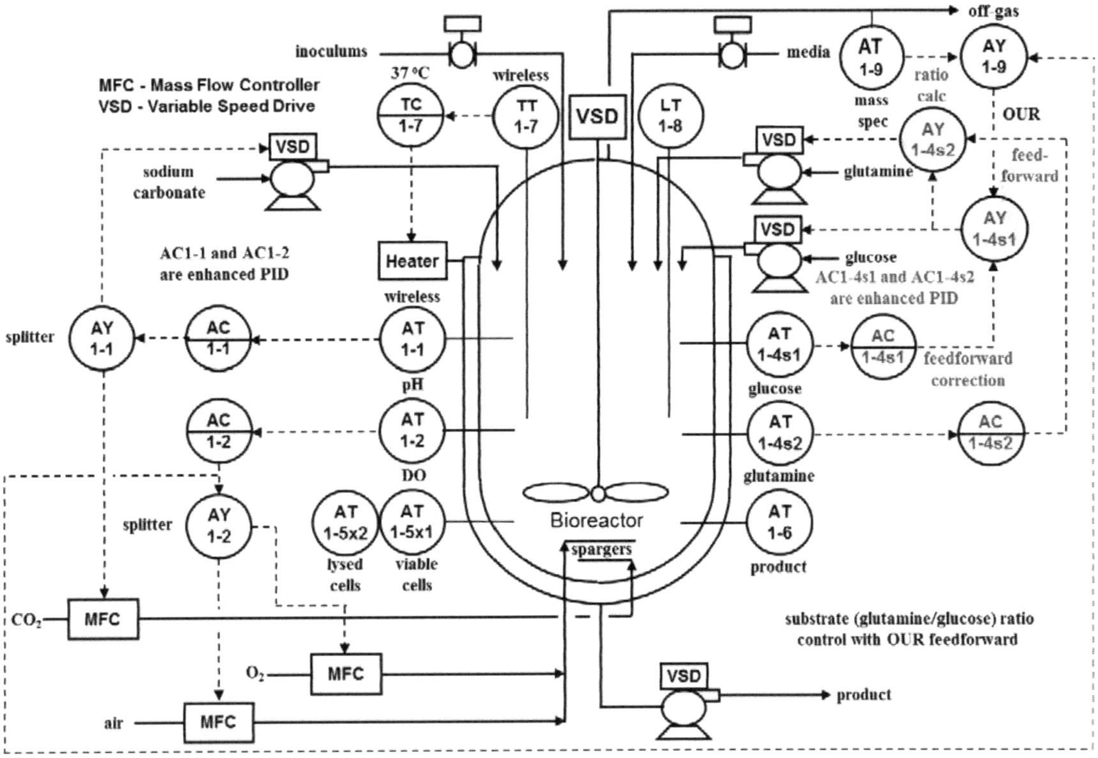

Figure 13 Glutamine to Glucose Ratio Control of Mammalian Bioreactor

New online and at-line analyzers for measuring glucose and glutamine open the opportunity for improved concentration control of this sugar and this amino acid. At present, glucose and glutamine addition are scheduled in anticipation of utilization rates. A move from sequenced batch charges to fed-batch closed loop control can compensate for disturbances and consistently maintain a concentration (Figure 13). The glutamine feed setpoint can be ratioed to the glucose feed setpoint. A feedforward signal for glucose addition rate based on oxygen uptake rate (OUR) can provide preemptive correction for the acceleration of substrate utilization in the exponential growth phase. The sampling rate for at-line analyzers doesn't need to be more frequent than once per 4 hours because the changes in mammalian cell metabolism are slow. An enhanced PID can provide tight control for large sample times and can suppress reaction to noise and feedforward timing errors by a threshold sensitivity setting.

Insight: Online and at-line analyzers of glucose and glutamine concentration offer the opportunity for optimizing the ratio of these concentrations for better cell performance (higher cell growth rate and product formation rate and better product quality).

Optimization of Mammalian Bioreactor Batch Profiles

If the rate of change is computed for viable cell and product concentration by passing these concentrations through a dead time block to create an old value that is subtracted from the new value and divided by a dead time set to provide a good signal to noise ratio, the growth rate and product formation rate can be indicated and controlled (Figure 14). The concentrations can be measurements from online or at-line analyzers or inferential measurements from OUR or a virtual plant running in synch with the actual plant.

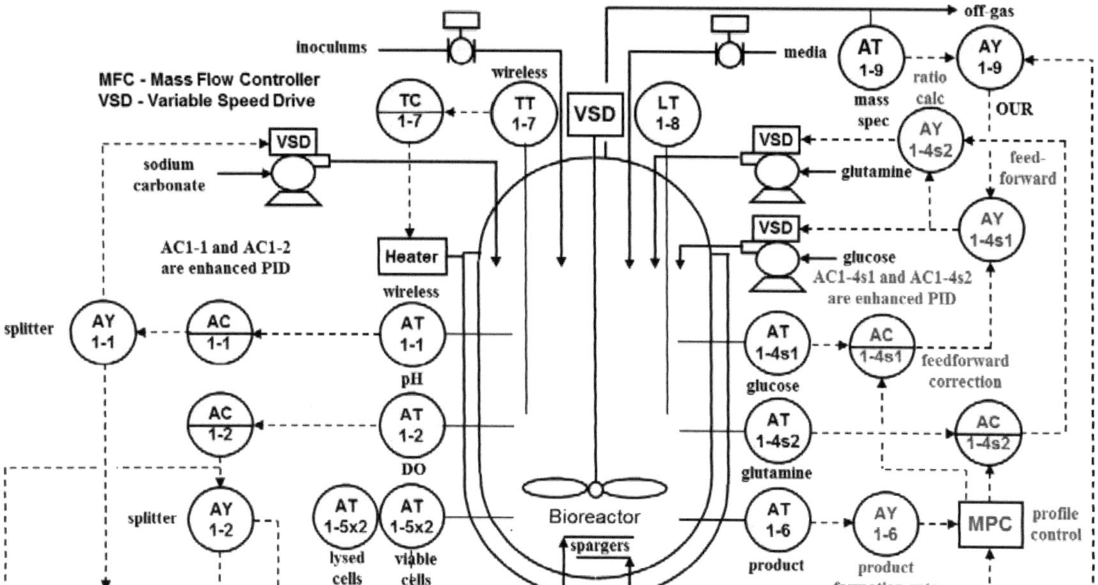

Figure 14 Optimization of Mammalian Bioreactor Batch Profiles

Profile control can provide a more consistent and faster rate (e.g., cell growth rate and product formation rate) at the most appropriate time. The glucose setpoint or the glutamine to glucose ratio may be directly manipulated by cell growth rate or product formation rate controllers whose setpoint is a function of batch phase. Product formation rate control may start later in the batch at which point growth rate control may be suspended or given less priority through tuning. An enhanced PID is used for these controllers to deal with discontinuous measurements, noise, and interactions.

Insight: An enhanced PID, and a dead time block computation of the rate of change of a concentration, create the opportunity for batch profile slope control to provide a more consistent and optimized cell growth rate and product formation rate.

The manipulation of temperature or pH as a secondary control loop for profile control requires a translation of the controlled variable from rate of change (velocity) to acceleration and deceleration to deal with the peak in the plot of growth rate or formation rate versus temperature and pH. The change in sign of the slope and process gain would confuse a controller whose process variable is the slope. The creation of a new controlled variable for the secondary loop that is the change in batch profile slope with respect to a change in temperature or pH eliminates the reversal of the sign of the process gain. The controlled variable decreases as the operating point approaches the optimum in the plot of cell growth rate or product formation rate versus pH or temperature. The controlled variable is zero at the peak and becomes more negative as the operating point moves further to the right of the peak.

The manipulated variable for this new secondary controller is the temperature or pH setpoint of a tertiary lower loop. Thus, there is a triple cascade of batch profile slope to batch profile acceleration to temperature or pH. However, unless Model Predictive Control (MPC) is used, the manipulation of either temperature or pH may be done and not both, due to interactions. In addition, careful setpoint limits must be placed on the tertiary controller. If PID controllers are used, external reset feedback throughout the triple cascade is critical to prevent an upper PID output from changing faster than a lower PID process variable can respond. External reset feedback helps deal with the violation of the cascade rule caused by the secondary loop not being 5 times faster than the primary loop. The enhanced PID is advisable to suppress oscillations and tuning problems posed by the discontinuous updates from analyzers.

The triple cascade is an extension of the "Hill Climbing" strategy described by Shinskey in *Controlling Multivariable Processes*. To create the innovative secondary loop, the new controlled variable is scaled from a positive change in velocity (acceleration) to a negative change in velocity (deceleration) with respect to a change in temperature or pH. For a constant profile slope, the setpoint is zero. The acceleration and deceleration can be computed by passing the concentration rate of change through a dead time block to give a significant difference between a new and an old value of the slope to compute a second derivative of the change in concentration.

With this calculation, it is critical to choose a dead time interval that is long enough to provide a strong signal to noise ratio. The change in temperature or pH is first computed in the same manner by the use of a dead time block with the same dead time parameter. This change in temperature or pH must be dynamically compensated so that a change in profile is coincident with a change in temperature or pH. The dynamic compensation can be identified by the use of MPC software similar to what is done for linear dynamic estimators except that the time intervals between step tests must be larger than the dead time block parameter used to create the rate of change signals. The calculation must be protected against a divide by zero.

When the triple cascade is turned on, a change in the manipulated temperature or pH in the most desirable direction should be initiated to create a change in slope. The tuning of the primary PID must be slower than that of the secondary PID, which must be slower than that of the tertiary PID. Since this is an optimization, and biological batch cycle times are extremely long, patient, gradual corrections are in order by the use of exceptionally large reset times and low PID gains. The product of the PID gain and reset time must be greater than four times the inverse of the integrating process gain for the primary and secondary loops to prevent the start of slow rolling oscillations from violation of Equation (7).

Smooth and gradual action is also important from the standpoint of not reacting to noise. The numerator used in the calculation of the controlled variable for the secondary PID is a second derivative of the batch profile with respect to time giving the acceleration or deceleration in the batch profile with respect to a change in pH or temperature. The computation is consequently susceptible to noise.

Insight: For maximizing the cell growth rate or product formation rate, a triple cascade extension of the Shinskey "Hill Climbing" technique can optimize the pH or temperature setpoint to be near the peak in the plot of cell growth rate or product formation rate versus pH or temperature. For temperature, the target is just to the left of the peak because of the risk of cell death from temperatures to the right of the peak from sensor offset or setpoint overshoot.

Nomenclature

E_i = integrated error from load disturbance (e.u. $*$ sec)

E_o = open loop error (load disturbance error with controller in manual) (e.u.)

E_x = peak error from load disturbance (e.u.)

K_c = controller gain (dimensionless)

$K_{c\min}$ = minimum controller gain (dimensionless)

K_i = open loop integrating process gain (1/sec)

K_o = open loop self-regulating process (steady state) gain (%/%) (dimensionless)

K_o' = open loop runaway process gain (dimensionless)

K_x = stability factor (oscillations start at 4 & oscillations sustained at ¼) (dimensionless)

R_c = ratio of maximum to minimum controller gain (dimensionless)

R_h = ratio of heat removal to reactor positive feedback time constant (dimensionless)

R_m = ratio of measurement to reactor positive feedback time constant (dimensionless)

T_i = controller integral time (reset time) (sec)

T_d = controller derivative time (rate time) (sec)

λ = Lambda (closed loop time constant or arrest time) (sec)

θ_o = total loop dead time (sec)

τ_f = signal filter time constant (sec)

Δt_x = controller execution time (PID execution rate) (sec)

τ_o = open loop negative feedback time constant (largest time constant in loop) (sec)

τ_o' = open loop positive feedback time constant (largest time constant in loop) (sec)

τ_s = secondary time constant (second largest time constant in loop) (sec)

τ_p = negative feedback process time constant (largest time constant in loop) (sec)

τ_p' = positive feedback process time constant (largest time constant in loop) (sec)

Conclusion

The best reactor performance is achieved by the knowledge of reactor and sensor dynamics leading to good measurements, tuning, and control strategies. The use of resistance temperature detectors (RTDs) and Coriolis meters offer accurate drift fee measurements. Except for some plug flow reactors, integrating process tuning rules provide the best loop performance for both batch and continuous reactors. Control strategies to tightly control stoichiometry and maximize yield can be set based on the reactant and product phases. The addition of a valve position controller can maximize capacity. Except for level and residence time control, the strategies for continuous reactors can be applied to fed-batch reactors. For a unidirectional response due to a non-reversible reaction and/or absence of opposing manipulated variables (e.g., heating without cooling, acid without base), a simple calculation of the slope of the batch profile can be used as a controlled variable for profile control and prediction of end point.

References

Harriott, P. *Process Control* McGraw-Hill, 1964

Luyben, W. and Cheung, T. "Proportional-Derivative Control of Exothermic Reactors" Joint Automatic Control Conference, Philadelphia, Vol. 3, October, 1978

Luyben, W. and Melcic, M. "Consider Reactor Control Lags" *Hydrocarbon Processing,* Vol. 57, No. 3, 1978

Luyben, W. and Tyreus, B. "Dynamics and Control of Recycle Systems - Ternary Systems with One or Two Recycle Streams" Ind. Eng. Chem. Res, Vol.32, No. 6, 1993

Luyben, W. and Luyben M. *Essentials of Process Control* McGraw Hill, 1997

Luyben, W. *Chemical Reactor Design and Control* Wiley, 2007
McMillan, G and Vegas, H. *101 Tips for a Successful Automation Career* ISA, 2014
McMillan, G. *Advances in Reactor Measurement and Control* ISA, 2015
McMillan, G. *Good Tuning: A Pocket Guide* – 4th Ed. ISA, 2015
McMillan, G. *Tuning and Control Loop Performance* – 4th Ed. Momentum Press, 2015
Shinskey, G. *Energy Conservation Through Control* McGraw-Hill, 1978
Shinskey, G. *Controlling Multivariable Processes* ISA, 1981
Shinskey, G. *Feedback Controllers for the Process Industries* McGraw-Hill, 1994
Shinskey, G. *Process Control Systems* 4th Ed. McGraw-Hill, 1996
Smith, C. *Practical Process Control – Tuning and Troubleshooting* Wiley, 2009
Smuts, J. *Process Control for Practitioners* OptiControls, 2011.

Biography

Greg is a retired Senior Fellow from Solutia Inc. During his 33 year career with Monsanto Company and its spin off Solutia Inc, he specialized in improving loop performance, controller tuning, valve dynamics, opportunity assessments, dynamic simulation, fermenter control, pH control, and reactor control. Greg is the author of more than 100 articles and papers and 20 funny and serious books. His latest technical books are *101 Tips for a Successful Automation Career, Advances in Reactor Measurement and Control, and Tuning and Control Loop Performance – 4th Ed.* Greg has written a monthly "Control Talk" column for *Control* and has written 30 articles for *Control* and *InTech* since 2002. Greg posts Control Talk Blogs on the *Control* Global site, insights on an International Society of Automation (ISA) site, and video tutorials on the MYNAH Technology site.

Greg is an ISA Fellow and received the ISA "Kermit Fischer Environmental" Award for pH control in 1991, the *Control* "Engineer of the Year" Award for the Process Industry in 1994, and was one of the first inductees into the *Control* "Process Automation Hall of Fame" in 2001. Greg was honored as one of *InTech's* 50 most influential innovators for advancing automation and control technologies in 2003 and received the ISA Life Time Achievement Award in 2010. Greg started an ISA Mentor Program in 2012 for users in the process industry with 2 or more years of plant experience. Greg and Hunter Vegas serve as mentors with the help of industry experts.

Greg.McMillan@Mynah.com
https://www.mynah.com/knowledge-base/greg-mcmillans-series-dynamics
http://automation.isa.org/author/gregmcmillan/
http://www.controlglobal.com/blogs/controltalkblog/
http://www.controlglobal.com/voices/mcmillan-weiner/
http://modelingandcontrol.com/archives/deminars/

Advanced Control of Multiple Sulfur Units

Dustin Beebe
ProSys, Inc

15 AIChE
Spring Meeting, Austin, TX

PROSYS Maximize Your Plant.™

Presentation Agenda

- Brief Introduction of ProSys
- Wild Loads Overview
- Control Issues
- Control Solutions
- Benefits

15 AIChE
Spring Meeting, Austin, TX

PROSYS Maximize Your Plant.™

Engineering Services

- Engineering Services include:
  - Basic Control
  - Advanced Control
  - Alarm Management
  - Operator Interface
  - Dynamic Simulation
  - Many More…

15 AIChE Spring Meeting, Austin, TX

PROSYS Maximize Your Plant.

Presentation Overview

- A control scheme to control multiple units receiving wild feeds from multiple sources or providing a utility or product to multiple users
- Developed using standard PID controllers and custom programming within the control system
- This is an example of Advanced Regulatory Control (ARC) – no predictive process model
- This control scheme installed in multiple industrial locations

15 AIChE Spring Meeting, Austin, TX

PROSYS Maximize Your Plant.

What is a Wild Load?

- A wild load is a supply or demand to/from multiple locations external to a group of units that must deal with it
- May be controlled by user or producer
- The wild load is not controllable by the unit which must receive or supply the stream
  - Must respond to unpredictable and sometimes large, rapid changes in demand
  - Usually little or no surge capacity

15 AIChE Spring Meeting, Austin, TX

PROSYS Maximize Your Plant™

Wild Load Handler Examples

- Producers
  - Boilers / Steam distribution
  - Hydrogen plants
  - Syngas plants
  - Air compressors
- Acceptors
  - Acid gas handling
  - Waste gas incinerators
  - Wastewater treatment
- We will call these Wild Load Handlers (WLH)

15 AIChE Spring Meeting, Austin, TX

PROSYS Maximize Your Plant™

Control Issues

- When multiple WLH in parallel are used to handle feed from or supply to multiple sources, the control system must:
  - Respond to uncontrolled load changes
  - Ensure that the wild load is properly balanced so that one WLH is not overloaded while another runs well below its capacity
  - React rapidly to one WLH trip, raising the loads on the others to maintain stability

15 AIChE
Spring Meeting, Austin, TX

PROSYS Maximize Your Plant™

Consequences of Control Failure

- Plant upsets in the units that use or generate the load
- May need to initiate steam shedding
- May need to reduce production
- May need to flare sour gas
- May lose control with inadequate instrument air
- Lost production / environmental exceedances

15 AIChE
Spring Meeting, Austin, TX

PROSYS Maximize Your Plant™

Typical Approach

- Typical approach to the control problem is to fix the load on most WLHs, and use only one as the swing unit
- Disadvantages
  - Lack of capacity to react to load changes
  - WLH loads not balanced
  - If one WHL trips, taking control of others is difficult, and requires substantial operator intervention

15 AIChE
Spring Meeting, Austin, TX

Control System Overview

- Load Balancing Advanced Regulatory Control (ARC)
  - Monitors header pressure (or total flow) and loads on each WLH
  - Adjusts target feed or product rates for each WLH
  - Supplemental ARCs adjust other key variables within each WLH in response to load and composition changes

15 AIChE
Spring Meeting, Austin, TX

PROSYS Maximize Your Plant.™

Control System Overview

- Load Balancing ARC
  - Master controller maintains gas header pressure
  - Pressure cascades to a total flow controller
  - Slave controllers maintain feeds to each WLH
  - All flows are P/T compensated

15 AIChE Spring Meeting, Austin, TX

PROSYS Maximize Your Plant™

Control System Overview

- Load Balancing ARC
  - Constraints included for each WLH as necessary
    - Temperature / pressure limits within each WLH
    - Equipment capacity
    - Valve saturation

15 AIChE Spring Meeting, Austin, TX

PROSYS Maximize Your Plant™

Load Balancing

Sulfur Recovery Units Example

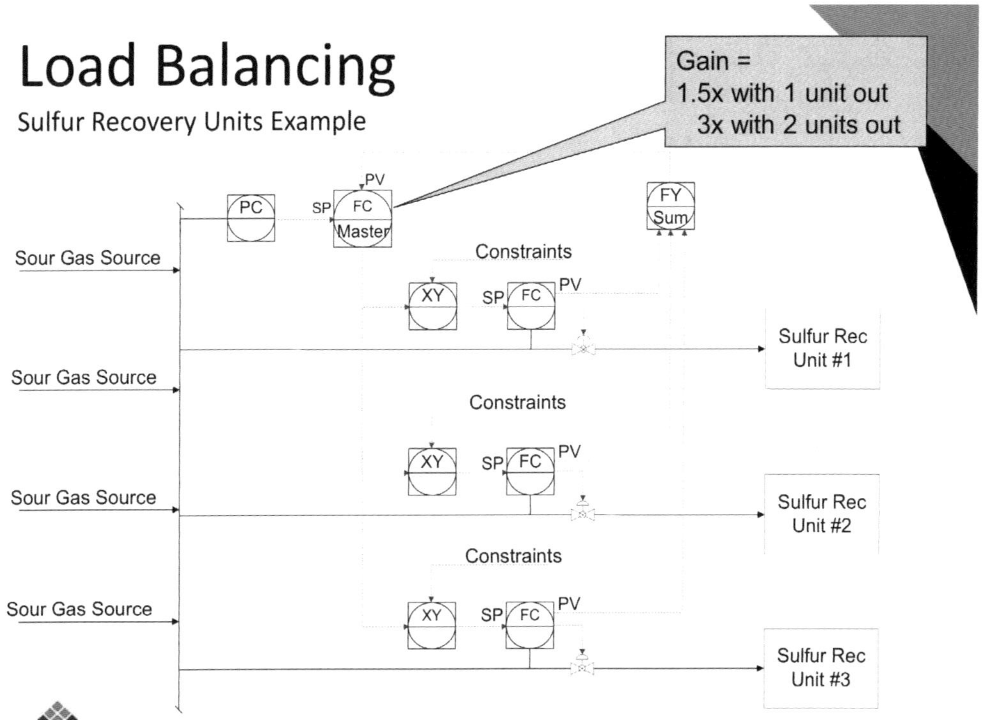

15 AIChE Spring Meeting, Austin, TX

PROSYS Maximize Your Plant.

Control System Overview

- Adaptive Gains
  - Adaptive gains employed in the master flow controller to improve header pressure control
  - Based on number of WLHs available (not at constraint), gains on setpoints of flow controllers in other WLHs adjusted
  - Examples
    - If 3 WLHs installed in parallel, only 2 are available, gains to the available WLHs must be 3/2 x the base value
    - If 4 installed, 3 available, gains are 4/3 x

15 AIChE Spring Meeting, Austin, TX

PROSYS Maximize Your Plant.

Load Balancing

Boilers Example

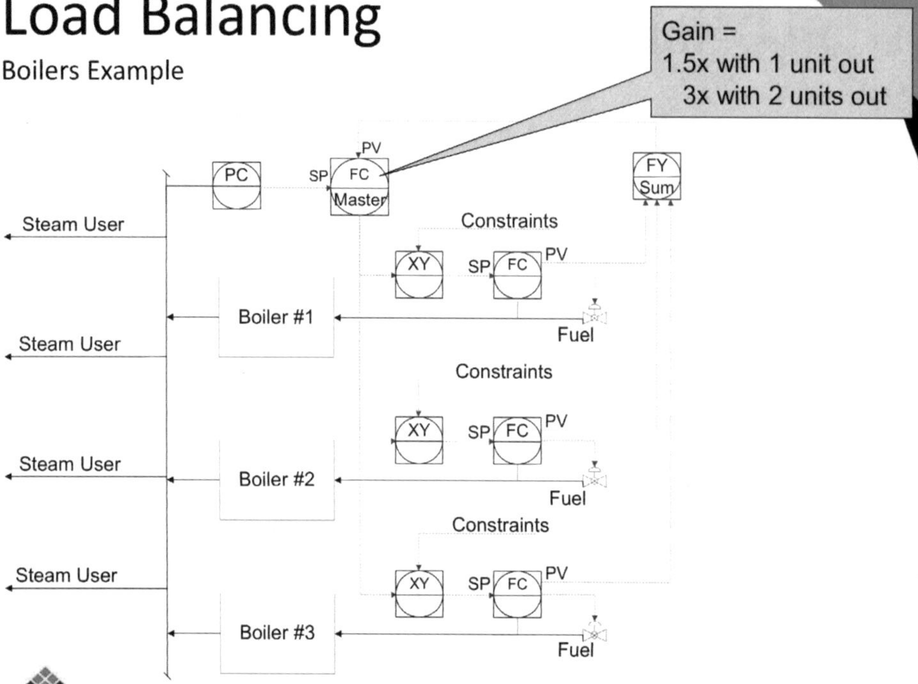

15 AIChE Spring Meeting, Austin, TX

PROSYS Maximize Your Plant.

Constraint Control

Boiler Example

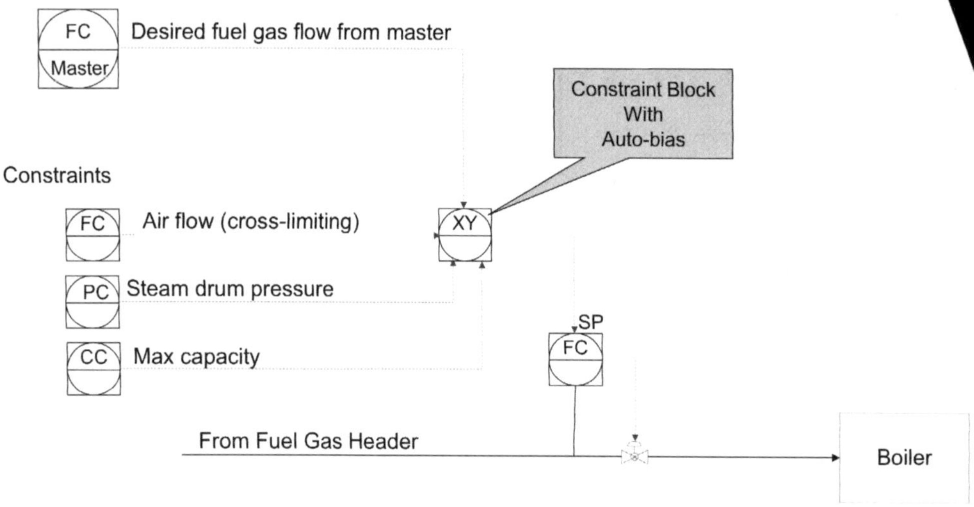

15 AIChE Spring Meeting, Austin, TX

PROSYS Maximize Your Plant.

Supplemental ARCs

- Within each WHL, individual controls must also deal with changing loads
- Must be considered with load-balancing design
- Details of ARCs that accomplish this are beyond the scope of this presentation
- Examples of supplemental ARCs
  - Air/O2 balancing to SRU burners
  - Fuel/Air cross-limiting controls
  - Feeds ratio control
  - Feed forward furnace controls
  - BFW controls (3 element)

15 AIChE
Spring Meeting, Austin, TX

PROSYS Maximize Your Plant.

Handling Control Issues

- Respond to uncontrolled load changes
- Ensure that the wild load is properly balanced so that one WLH is not overloaded while another runs well below its capacity
- React rapidly to a WLH trip, raising the loads on the others to maintain refinery stability

Load Balancing ARC with Adaptive Gain

15 AIChE
Spring Meeting, Austin, TX

PROSYS Maximize Your Plant.

Benefits

- Improved control – fewer impacts to users or upstream units
- Fewer issues when a trip occurs
  - No need to adjust production / operation in user plants
  - Keep production units on line
- Better control within WLH

15 AIChE Spring Meeting, Austin, TX

PROSYS Maximize Your Plant.™

Benefits

- Little or no operator intervention unless all units are at constraint limit
- Handled by ARC, without model-predictive control (MPC)
  - No external MPC application
  - No process model required
  - Process need not be linear

15 AIChE Spring Meeting, Austin, TX

PROSYS Maximize Your Plant.™

Before / After – Case Study

- Refinery fighting control issues in sour gas header
  - Before ARC installed, multiple operating incidents and several permit violations over year's time due to ineffective controls
  - After ARC, no violations in a full year
  - Project cost recovered within a few months
  - Can now withstand sudden loss of a sulfur unit without incident

15 AIChE Spring Meeting, Austin, TX

PROSYS Maximize Your Plant.™

Questions?

Comments?

15 AIChE Spring Meeting, Austin, TX

PROSYS Maximize Your Plant.™

Advances in Developing New MPC & Advanced Process Control Solutions

Name – Maina Macharia
Title – Process Technical Consultant
Co – Rockwell Automation

Date – April 29, 2015

PUBLIC

Connect with us.
www.rockwellautomation.com www.us.endress.com

15 AIChE
Spring Meeting, Austin, TX

PUBLIC

MPC – Value Proposition

Reduce Process Variance

1. Increase Plant Stability
2. Increase Feedstock yield – Reduce waste
3. Reduced Energy of production
4. Increased production (efficient asset utilization)
5. Improved Quality control

Increased Margin and Profitability

PUBLIC

How MPC Generates Benefits

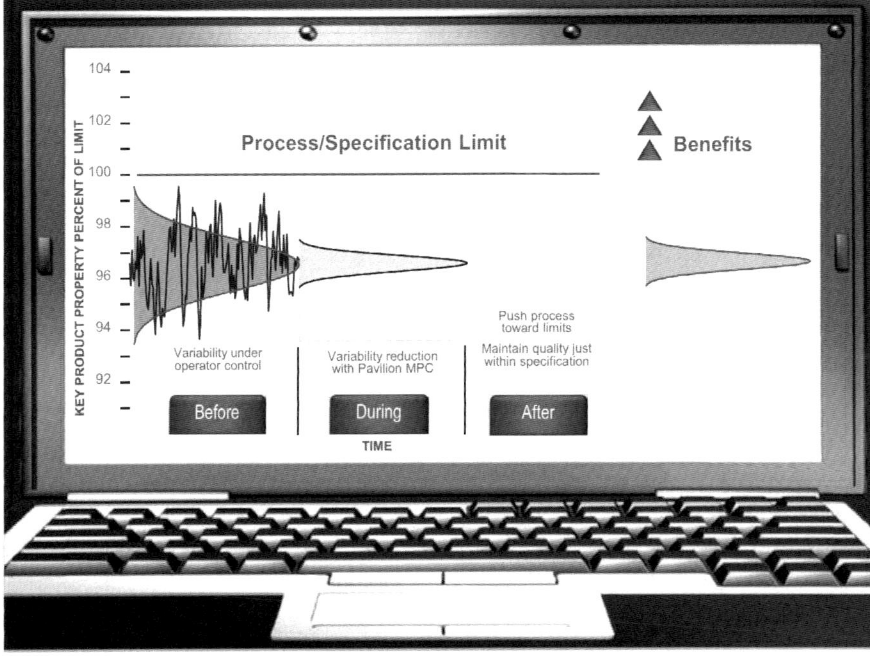

REDUCES Variability

ACHIEVES "Plant Obedience"

MANAGES the process within constraints

Capacity Utilization– operate closer to specifications and performance limits while maintaining safety margins

3

$490 Million Global MPC shipments-Regions – (2012)

Total Shipments of Advanced Process Control & On-Line Optimization by World Region

Figures in Millions of US Dollars

World Region	2012	2013	2014	2015	2016	2017	CAGR
North America	178.2	184.4	192.6	202.7	215.8	229.7	5.2%
EMEA	169.4	174.7	181.8	190.4	201.6	213.4	4.7%
Asia	102.1	107.1	113.8	122.3	133.4	145.5	7.4%
Latin America	37.5	39.1	41.3	44.0	47.6	51.5	6.6%
Total	487.1	505.3	529.5	559.5	598.3	640.0	5.6%

Source: ARC MPC Report 2012

PUBLIC 4

Multivariable Predictive Control : Structure

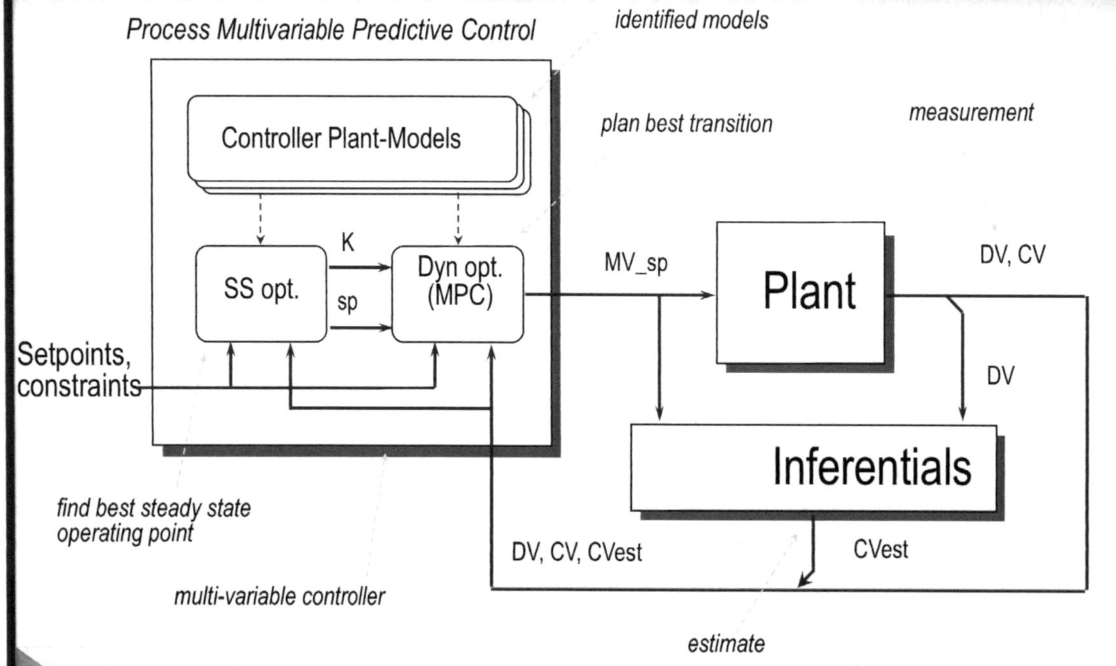

PUBLIC

Dynamic Optimization Function and Constraints

- CV Desired
- CV Maximum Limit
- CV Minimum Limit
- CV Upper Convergence cone
- CV Lower Convergence cone

- MV Desired
- MV Move suppression

Minimized objective function subject to constraints

- MV Rate of Change Constraints
- MV Hard Constraints

PUBLIC

Minimization of MPC Objective Function

Minimize objective function, subject to a set of MV Constraints. Solution is always feasible

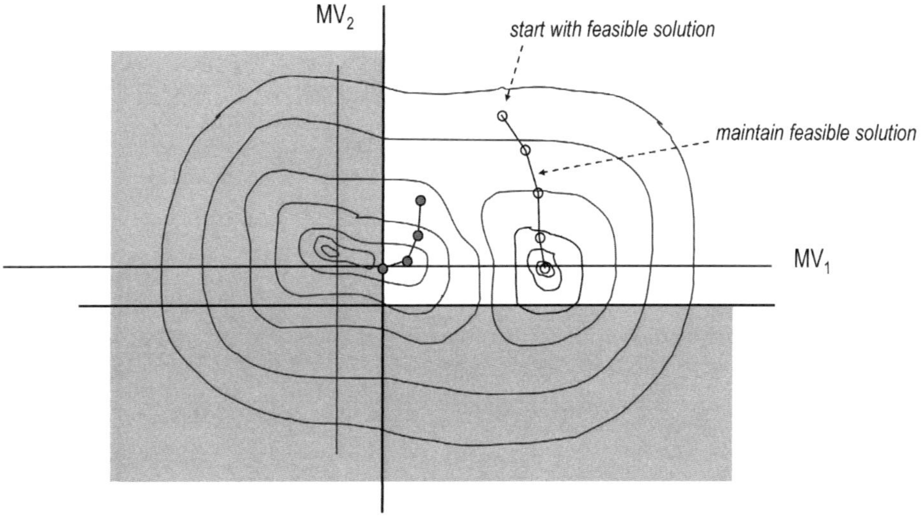

PUBLIC

Operator Interaction with MPC

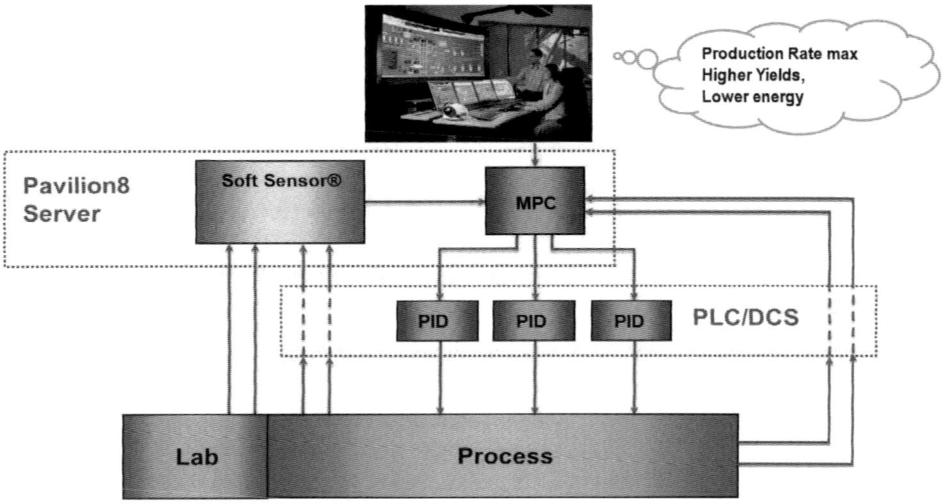

PUBLIC 8

Foundation –MPC Solutions

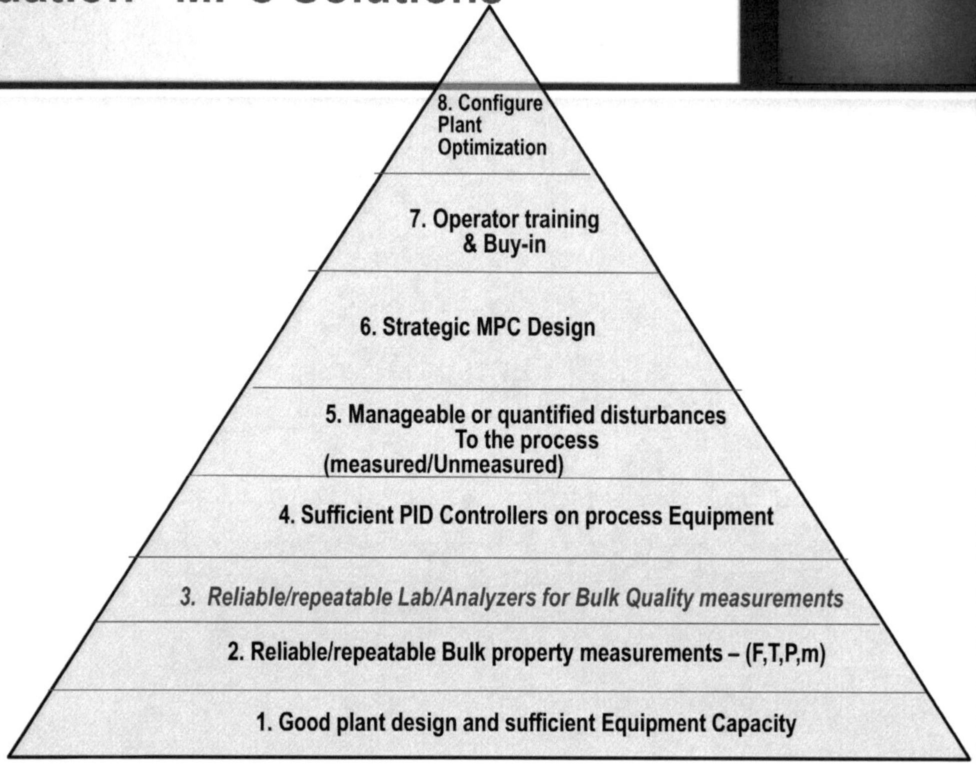

PUBLIC

Productivity: Automatic MPC Tuning

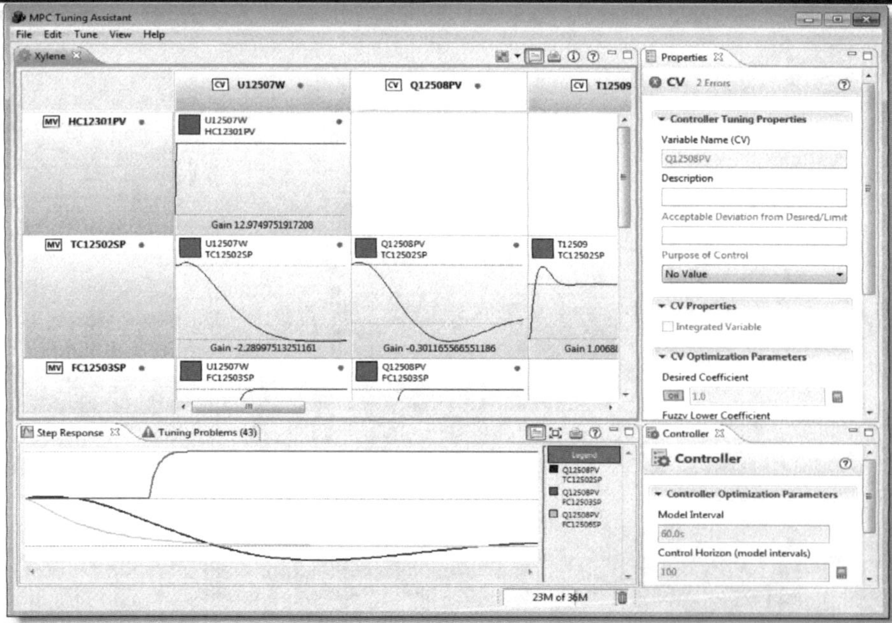

Simplifying the process of tuning MPC controllers

PUBLIC

Typical Project KPI – Monitor& Tracking

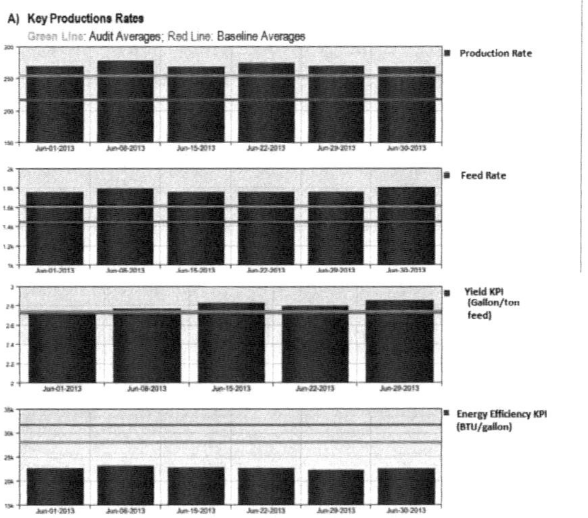

MPC – Performance Metrics

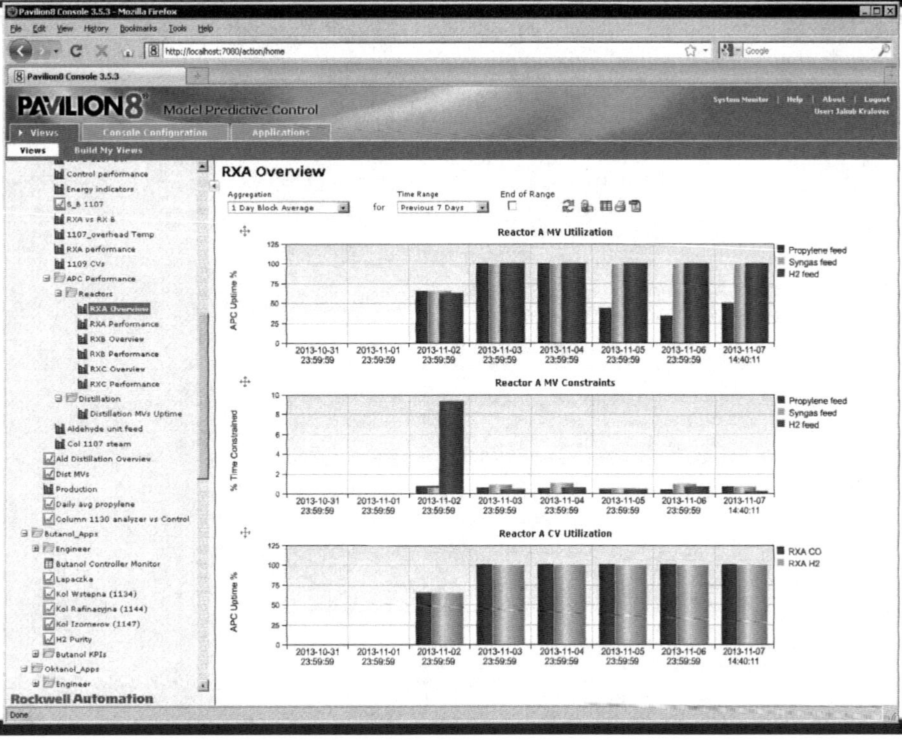

New Applications Areas

1. Semi-Continuous process with Start/Stop
2. Non-linear Model Predictive Control Applications
3. End-point Model Predictive Control Applications
4. Batch Model Predictive Control Applications

Expanded MPC Applications

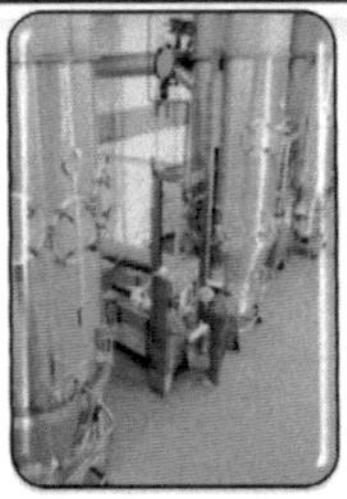

Start/Stop Process
- Spray Dryers
- Evaporators
- Energy Centers
- Crystallizers
-

Process Types
- Milk Powder
- Coffee
- Laundry Detergent
- Conc. Juice
- Sugar

End-point Control

Typical Benefits
- 2 to 5% production increase
- 2 to 5% energy consumption reduction
- 20 to 40% product variability reduction
- 10 to 30% off-spec product reduction

Non-Linear MPC

Process Types

Typical Benefits
- 4 to 8% prime product yield increase
- 35 to 75% product variability reduction
- 20-40% transition time reduction
- 3 to 7% feed stock wastage reduction

Batch MPC

Typical Benefits
- 4 to12% ethanol production capacity increase
- 2 to 5% ethanol yield increase
- 3 to 6% energy use/gallon reduction
- 1 to 2% DDGS yield increase

Dryer Moisture Soft Sensor Results

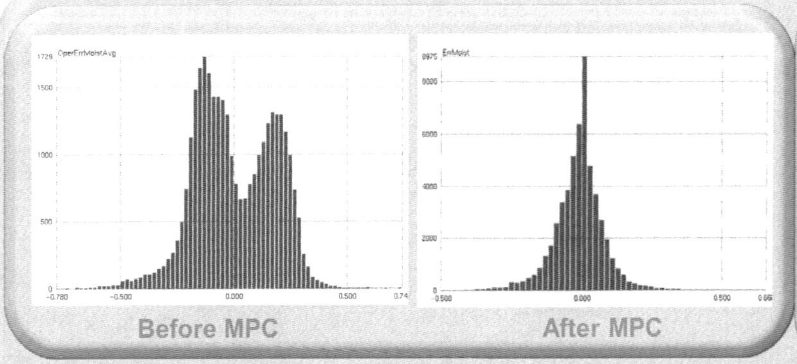

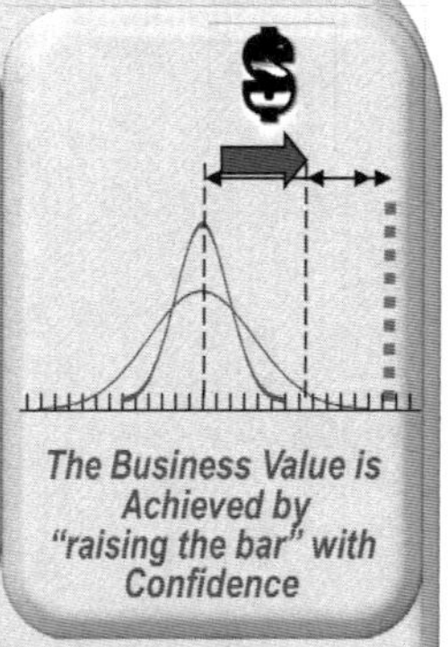

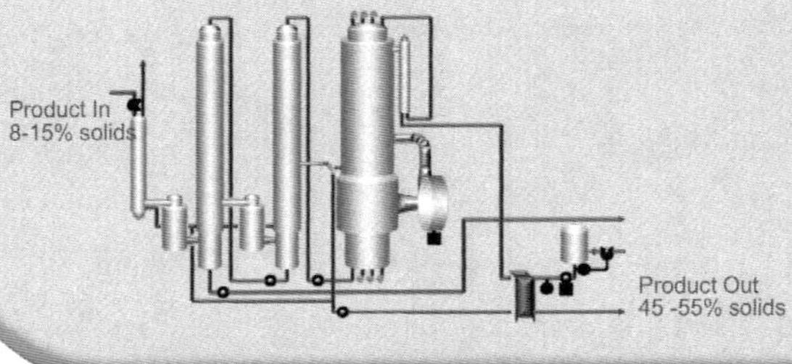

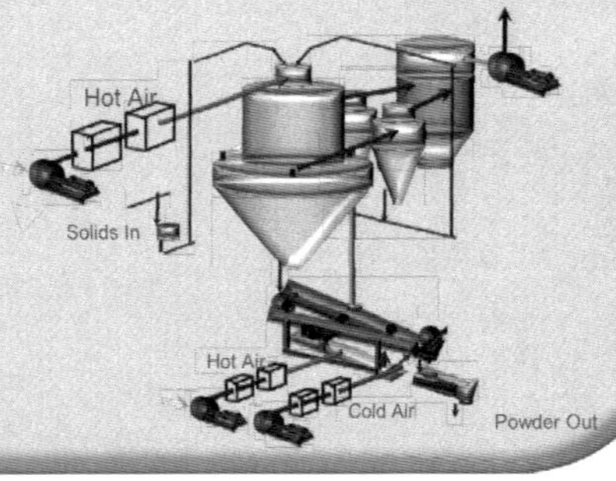

15

PUBLIC

Dryer and Evaporator Balance – Value

- Plant Obedience (reduced variability) enables:
- **Moisture** – Lift moisture targets to Limit or revised uplift
- **Evaporator capacity** – Optimize Solids to current UCL to reduce variability
- **Dryer capacity** – Improve evaporation capability within proven heat envelope
- **Reduce Rework** - Product & Cost
- **Reduce Energy** – Dryer & Evaporation Balancing reduces the energy cost profile per tonne of product produced .

- Improve Quality & Consistency
- Optimize Evaporator & Dryer Efficiency
  - Squeeze more value from the existing asset
  - Produce more product with the same operating profile
  - Manage powder glass transition
  - Enable product mix flexibility to gain more higher value products

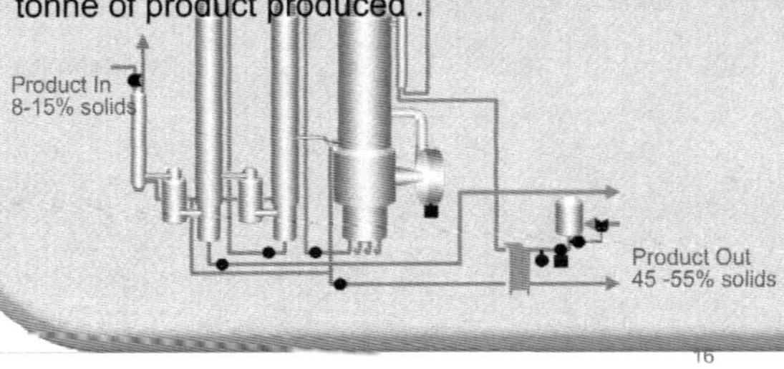

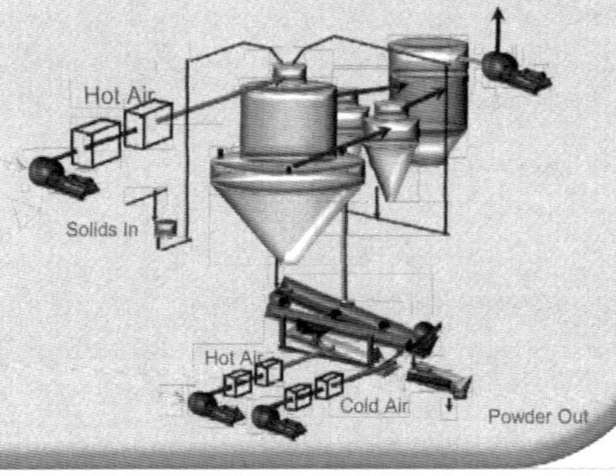

16

PUBLIC

Spray Dryer

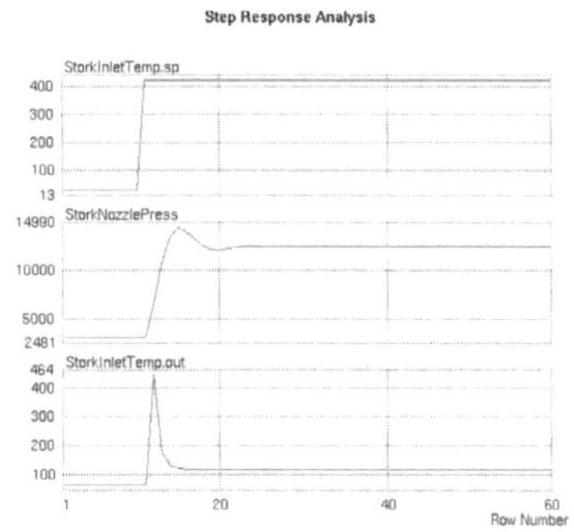

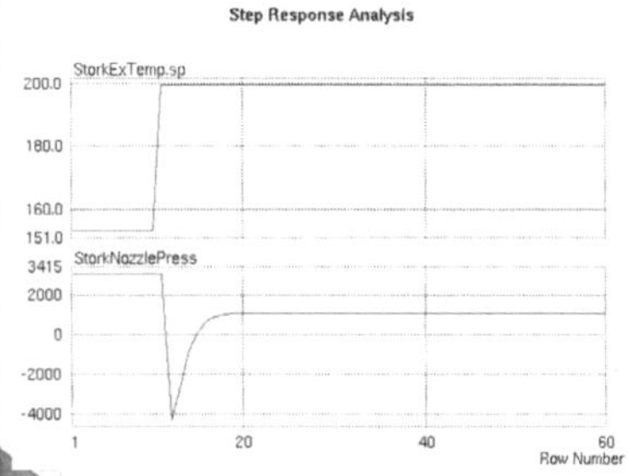

Spray dryer

- Inlet temp controlled with gas valve
- Exhaust temp controlled with feed pump speed
- Feed is pumped through nozzles and the backpressure from the nozzles is a key constraint
- Inlet temp sp to valve output, response of pid algorithm
- Inlet temp to nozzle press, response of exhaust temp pid algorithm rejecting disturbance from inlet temp change.
- Exhaust temp to nozzle press, response of exhaust temp pid algorithm. As with Inlet temp, omitting the overshoot will cause cycling and possibly trip the dryer.

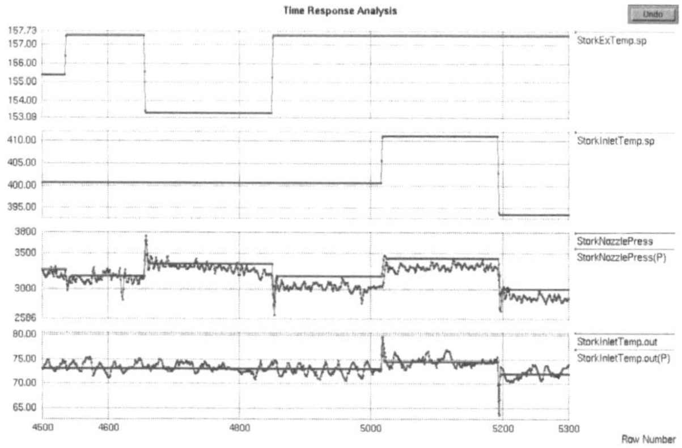

PUBLIC

Expanded MPC Applications

Start/Stop Process

Typical Benefits
- 5 to 8% production increase
- 30 to 60 % moisture variability reduction
- 20 to 50% off-spec product reduction
- 5 to 10% energy consumption reduction

End-point Control

Process Types
- French Fry Lines
- Mining conveyors
- Stockpile Blending

Process Types
- Food
- Mineral ore
- Minerals
- Polymers

Non-Linear MPC

Typical Benefits
- 4 to 8% prime product yield increase
- 35 to 75% product variability reduction
- 20-40% transition time reduction
- 3 to 7% feed stock wastage reduction

Batch MPC

Typical Benefits
- 4 to12% ethanol production capacity increase
- 2 to 5% ethanol yield increase
- 3 to 6% energy use/gallon reduction
- 1 to 2% DDGS yield increase

PUBLIC 18

Expanding MPC to Production Lines

PUBLIC 19

End-point Control

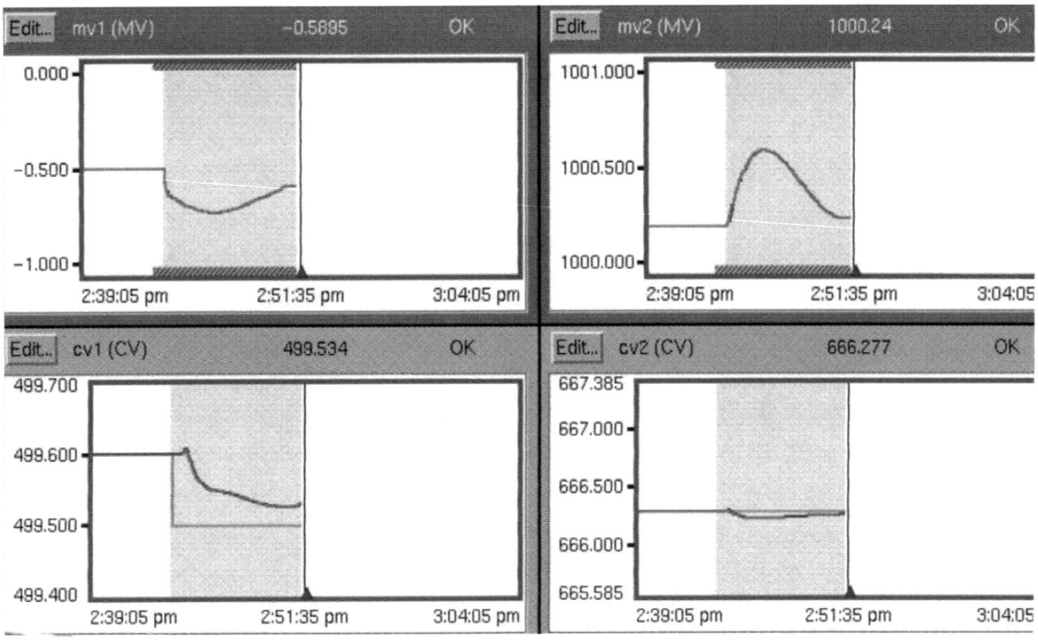

PUBLIC 20

Expanded MPC Applications

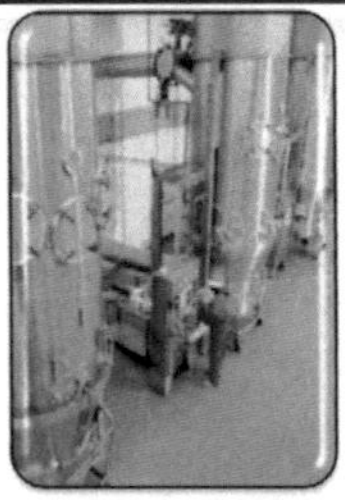

Start/Stop Process

Typical Benefits
- 5 to 8% production increase
- 30 to 60 % moisture variability reduction
- 20 to 50% off-spec product reduction
- 5 to 10% energy consumption reduction

Endpoint

Typical Benefits
- 2 to 5% production increase
- 2 to 5% energy consumption reduction
- 20 to 40% product variability reduction
- 10 to 30% off-spec product reduction

Non-Linear Control

Process Types
- **Polymers**
  - **PE, PP, PS, PC**
- **Ethylene Plants**
- **Styrene Plants**
- **Crude Refining**
- **Gas Plants**

Batch MPC

Typical Benefits
- **4 to12% ethanol production capacity increase**
- **2 to 5% ethanol yield increase**
- **3 to 6% energy use/gallon reduction**
- **1 to 2% DDGS yield increase**

PUBLIC 21

Before & After MPC Comparison

variables (PV-SP)	Before MPC STD	After MPC STD	MPC Improvemen
AIC201	12.23	4.48	63%
AIC202	148.5	46.03	69%
DIC241	2.540	1.29	49%
DIC251	2.402	1.58	34%
PIC402	0.01739	0.0032	82%
AIC402	0.00254	0.000296	88%
AIC403	0.00535	0.002977	44%

PUBLIC

AIC201/AIC202）H2 concentration in Loop 1&2 reactor

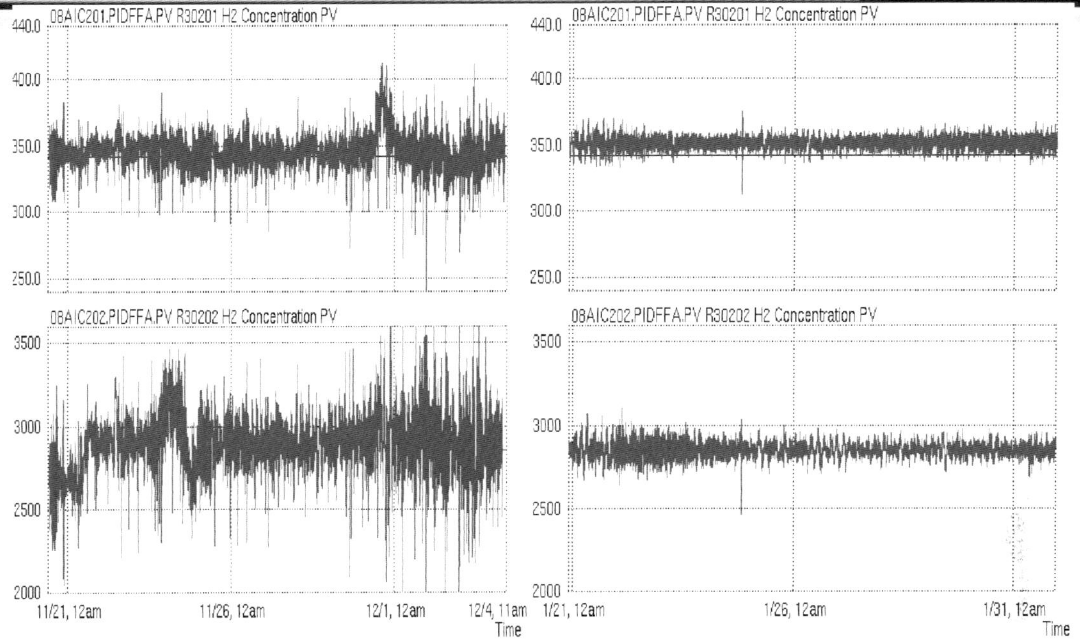

PUBLIC

Transition:V30G→Z30S, **MPC OFF** time: 7.5 hours

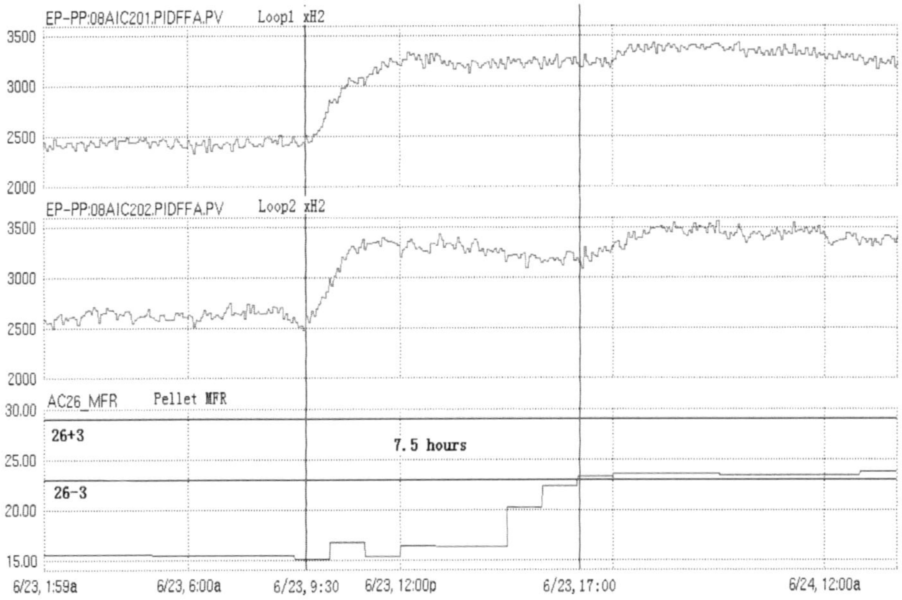

PUBLIC

Transition: V30G→Z30S, MPC ON time: 3.3 hours

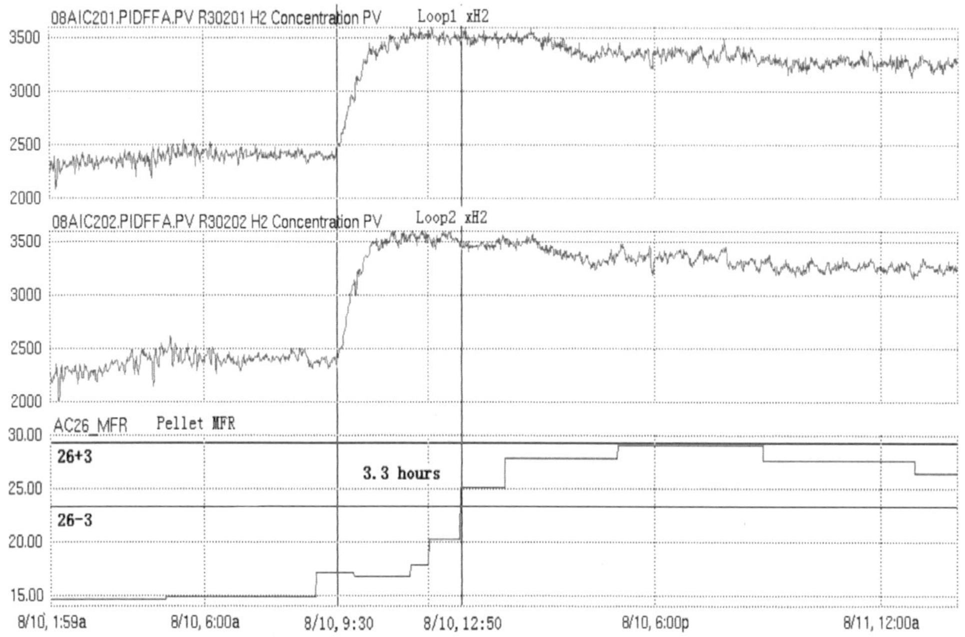

PUBLIC 25

Expanded MPC Applications

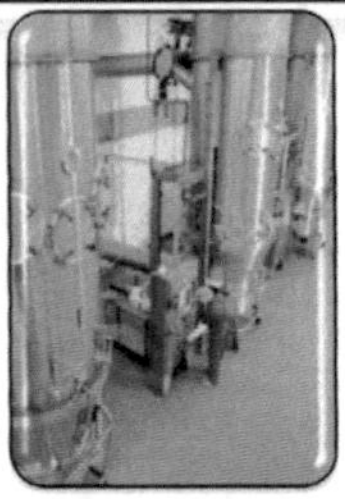

CPG

Typical Benefits
- 5 to 8% production increase
- 30 to 60 % moisture variability reduction
- 20 to 50% off-spec product reduction
- 5 to 10% energy consumption reduction

Endpoint

Typical Benefits
- 2 to 5% production increase
- 2 to 5% energy consumption reduction
- 20 to 40% product variability reduction
- 10 to 30% off-spec product reduction

Non-Linear Control

Process Types

Typical Benefits
- 4 to 8% prime product yield increase
- 35 to 75% product variability reduction
- 20-40% transition time reduction
- 3 to 7% feed stock wastage reduction

Batch MPC

- Fermentation

Process Types
- Corn Ethanol
- Cane Ethanol

PUBLIC 26

Bioreactor Control Challenges

- Unmeasured, infrequent, delayed process quantities (yeast, ethanol, sugar concentrations)
- Nonlinear process with complex interactions
- No steady state
- Online optimization

Develop an accurate, efficient process model

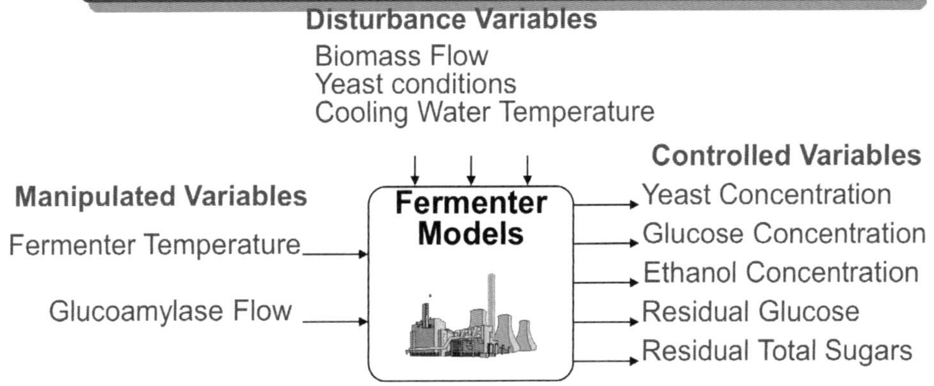

PUBLIC

Fermenters – Reduced Residual glucose

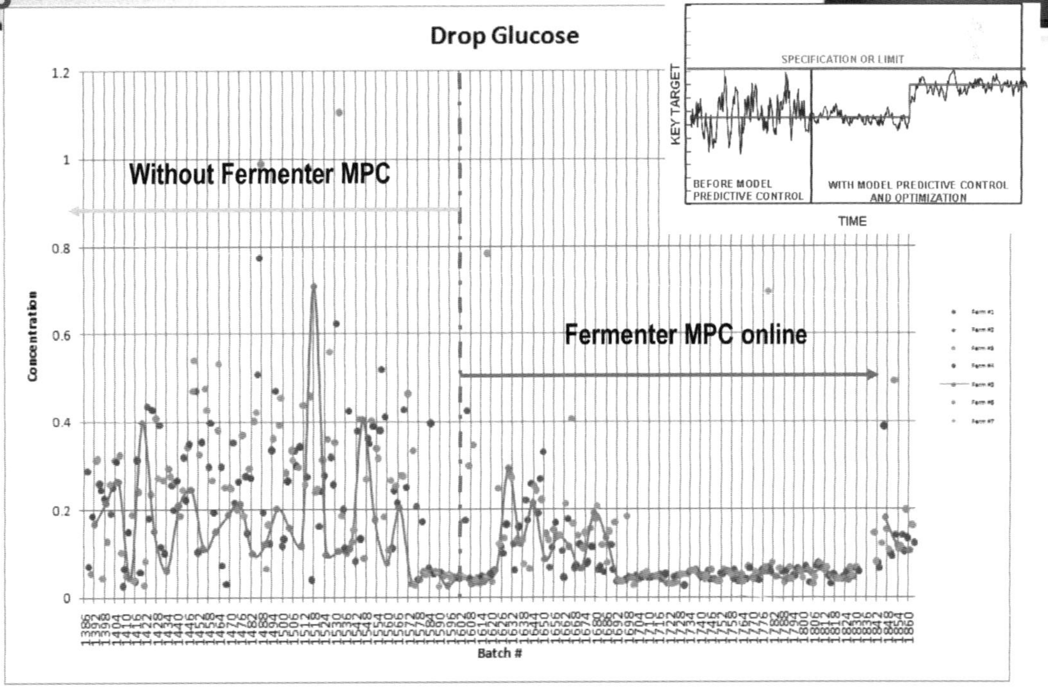

28

PUBLIC

Fermenters – Increased Ethanol Conc

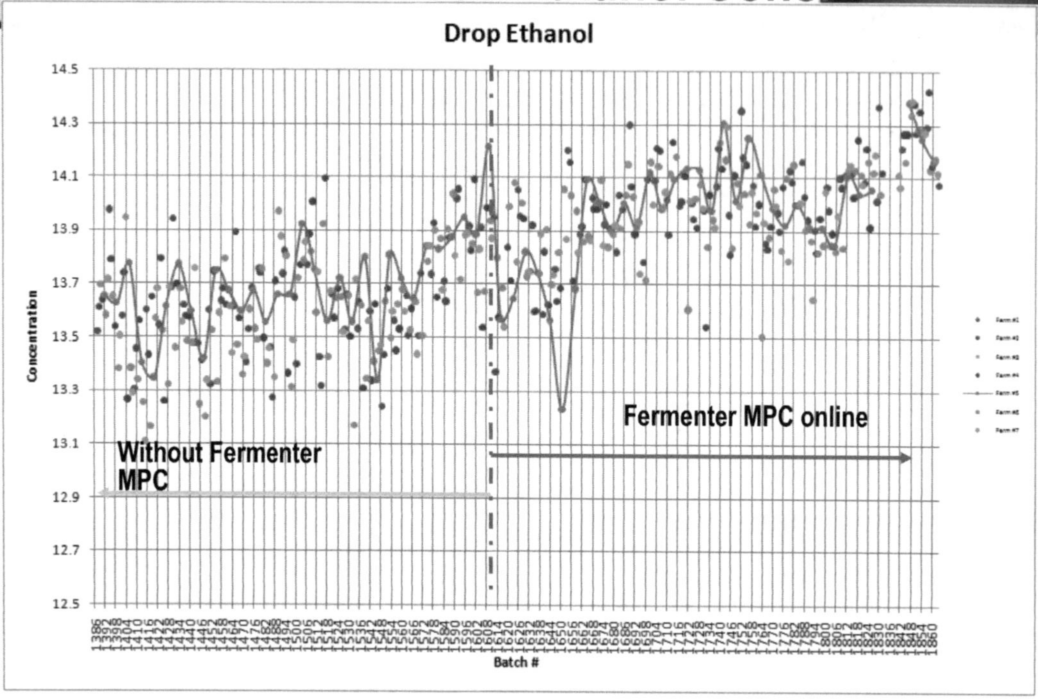

29

PUBLIC

Conclusion

- Significant Commercial Demand for new MPC industrial applications
- Ensure MPC improvements are tracked with KPI/graphs.These adds confidence to MPC Investments
- Traditional MPC has expanded outside Traditional distillation columns
  1. Start/Stops expanded MPC to non-continuous operations
  2. End-point is a conveyor efficiency enabler
  3. Non-linear control achieves high benefis
  4. Batch MPC together successfully implemented in Biofuels Industry
- Challenge for new applications is to be simple and adaptable for operators and maintenance staff

PUBLIC

30

Thank you!

PUBLIC

Connect with us.

www.rockwellautomation.com www.us.endress.com

PUBLIC

AUTHOR INDEX

American Institute of Chemical Engineers
3 Park Avenue
New York, New York 10016-5991

CURRAN ASSOCIATES INC
proceedings.com